# FIRE STANDARDS AND SAFETY

A symposium
presented at
National Bureau of Standards
Gaithersburg, Md., 5-6 April 1976

ASTM SPECIAL TECHNICAL PUBLICATION 614
A. F. Robertson, National Bureau of Standards, editor

04-614000-31

AMERICAN SOCIETY FOR TESTING AND MATERIALS
1916 Race Street, Philadelphia, Pa. 19130
NATIONAL BUREAU OF STANDARDS
Washington, D.C. 20234

75 YEARS
NBS
1901-1976

## NOTE

The Society is not responsible, as a body,
for the statements and opinions
advanced in this publication.

First Printing, Baltimore, Md.
March, 1977

Second Printing, Philadelphia, Pa.
February, 1986

# Foreword

The Symposium on Fire Standards and Safety was presented at a meeting held at the National Bureau of Standards, Gaithersburg, Md., 5-6 April 1976. The symposium was sponsored by The National Bureau of Standards and The American Society for Testing and Materials through its Committee E-5 on Fire Tests of Materials and Construction. A. F. Robertson, National Bureau of Standards, presided as symposium chairman.

# Related
# ASTM Publications

Ignition, Heat Release, and Noncombustibility of Materials, STP 502 (1972), 04-502000-31

# A Note of Appreciation to Reviewers

This publication is made possible by the authors and, also the unheralded efforts of the reviewers. This body of technical experts whose dedication, sacrifice of time and effort, and collective wisdom in reviewing the papers must be acknowledged. The quality level of ASTM publications is a direct function of their respected opinions. On behalf of ASTM we acknowledge with appreciation their contribution.

*ASTM Committee on Publications*

# Editorial Staff

Jane B. Wheeler, *Managing Editor*
Helen M. Hoersch, *Associate Editor*
Ellen J. McGlinchey, *Assistant Editor*
Kathleen P. Turner, *Assistant Editor*
Sheila G. Pulver, *Editorial Assistant*

# Contents

# Characterizing the Fire Hazard

*J. W. Lyons*[1]

# Fire Research and Fire Safety: A Status Report on the Situation in the United States

**REFERENCE:** Lyons, J. W., **"Fire Research and Fire Safety: A Status Report on the Situation in the United States,"** *Fire Standards and Safety, ASTM STP 614,* A. F. Robertson, Ed., American Society for Testing and Materials, 1977, pp. 5–10.

**ABSTRACT:** During the past century, efforts to control and reduce fire losses have concentrated successively on conflagrations destroying entire cities, large sections of cities, or large buildings, and, finally, on fires in the room of origin or in family residences. The first effort created effective fire departments, the second, the fire endurance concepts now an important part of the nation's building codes, and the third, now in full development, has produced research programs on interior finish, furnishings, and other consumer products. This current effort to protect occupants in the room of fire origin or family residences is proving to be the most challenging and complex, technically and socially, of the three. These occupancies vary in materials used, configuration of the rooms, and placement of an infinite variety of movable contents.

To assess the level of fire safety, we must be able to compute or measure the time-temperature and time-gas concentration profiles at a number of points in the room. This requires, first, careful full-scale research and concommitant theoretical modeling. Second, it requires smaller, more economical tests which are correlated with the full-scale work. A series of such smaller tests will be necessary, including smoke and gas evolution, rate of heat release, and rate of fire spread over surfaces. Work on all of these is active now in many laboratories. The symposium will provide interesting progress reports on some aspects of these current efforts.

**KEY WORDS:** fires, fire safety, research, buildings, flammability, furniture, fire tests, fire hazards

It is appropriate that fire research be the subject of the symposium published in this volume during the 75th anniversary observance at the National Bureau of Standards (NBS) and that it be jointly sponsored with the American Society for Testing and Materials (ASTM), for these two

[1] Director, Center for Fire Research, National Bureau of Standards, Washington, D. C. 20234.

organizations have been active in this field during most of their histories, and they have cooperated closely throughout this period. Today more than ever, ASTM and NBS are each coping with very serious challenges in the arena of fire safety. This symposium has been planned to illuminate some of the challenges, to indicate progress being made by many different groups, and to suggest when and how success may be achieved. This paper will describe the present difficulties and outline the direction we are taking.

Fire fighting has been a challenge to man probably ever since he moved out of the cave and certainly since he began to live in villages and towns. Surprisingly, fire fighting has been treated as a formal public responsibility only recently; for example, the London Fire Brigade began to receive tax support only in 1865.

Fire research in support of fire prevention and control has a distinguished history. Perhaps the first such work was reported for textile products in England in 1684. The French chemist Gay Lussac carried out extensive studies on means to improve fire resistance of fabrics used in theater curtains and in 1820 reported a list of substances he found effective on various fibers. William Henry Perkin, the great dye chemist, conducted research on flame proofing cotton fabrics and in 1913 published results which led to a commercial process. Gay Lussac and Perkin made most of the fundamental observations necessary to construct a science of chemical alteration of burning behavior by treating the condensed phase. For the most part, the rest of us have been ringing changes on their basic themes ever since.

Until the twentieth century, almost all efforts in fire control were directed to preventing holocausts, the loss of entire cities or large portions thereof. In the private sector, insurance companies sprang up to reduce the burdens of fire loss on individual property owners. The insurance underwriters were the driving force in creating fire companies. By about 1900, municipal fire services were well developed in most industrial countries, and, with a few notable exceptions, the disastrous urban fire was a thing of the past. By the 1920's, fires had been, in most instances, confined to the building of origin. To go further required research on building fires and some design innovations.

Fire researchers led by NBS's Ingberg proceeded much as we do today: they agreed on a most probable fire scenario, conducted full-scale fire tests, defined the critical control points, and suggested a means of using the results in controlling building design via the building codes. In this case, the relationship between fire load and fire severity was determined, and the concept of fire endurance requirements for structural components was developed. For a given amount of fuel in an enclosure, a fire of a certain intensity may be expected for a given length of time. Therefore one may, if one uses current knowledge and tools wisely, confine a fire to the area of origin and ensure that the building retains its structural integrity by designing the walls, floor and ceiling, and the structural members to resist the predicted fire exposure. Thus was developed the fire endurance concept known to us all in terms of ASTM Fire Tests of Building Construction and Materials (Including Tentative Revision) (E 119–75) and building code requirements for 1, 2, and 4-h ratings. During the half century since this work was undertaken, we have built our cities and towns so that, by and large, we expect fires to be contained within the room or area of origin. In fact, rarely do we lose an entire building which has been constructed to meet these fire endurance requirements. Most large buildings in the United States built during the period 1920 to 1960 are in this category.

In the last decade and a half, however, there has been an increasing use of new materials and new design concepts, the fire safety of which is not well enough understood. The research on which today's codes rest was done on traditional building materials (wood, gypsum plaster, concrete, and steel) used in certain traditional designs. Now we find many synthetic polymers used in buildings of more open design, and the old test methods and concepts are found wanting. Materials pass tests but fail in actual use. Hallowed procedures are suddenly found defective. There is confusion in the fire testing community.

We are entering a new era of fire safety design and enforcement in which at least three themes are receiving our attention: definition of the relative toxicity of combustion products from various materials, development of tests and standards that encompass the contribution of moveable furnishings to fire spread and growth, and modification and extension to residences of technology developed for large public and institutional buildings. Let us examine each of these in turn.

As attention shifted from protecting cities and entire large buildings to saving occupants in the area of fire origin and occupants in one- and two-family residences, concern developed about the effects of toxic combustion products. Data from studies sponsored by the Federal government and others showed that more people die from fires at home than anywhere else and that the direct cause of death is more often from inhaling toxic gases than from burns. The significance of this is more telling if one subtracts from the data fires involving apparel as the first item ignited. In other words, in building fires, smoke and toxic gases are the problem much of the time.

It has been the practice for authorities to put down as the cause of death in such cases "smoke inhalation." This is of no help if one is seeking ways to reduce the number of such fatalities. In the years since about 1970, several research programs have been undertaken to develop better understanding of the toxicity problem, with the ultimate goal of devising ways and means of mitigating its effects. Pathological tests of fire victims using techniques especially designed for the purpose have shed much light on the causes of these deaths. Carbon monoxide is the principle toxic agent but is often found in combination with alcohol, heart and circulatory disease, or other conditions producing exceptional vulnerability to stress. There is a significant fraction of cases—say 10 or 15 percent—which cannot be explained by either this mechanism or heat and flame. These fatalities have been attributed to toxic products more or less specific to the material burned, for example, hydrogen cyanide from nitrogen-containing materials, hydrogen chloride or other chlorine-containing gases from chlorine-based polymers, and so on. There are research programs active today to define the toxicity of combustion products from known materials. There have been some surprises; there will be more. We have learned that simplistic approaches will not do; that merely analyzing the gases for major components will miss the unexpected compound formed during combustion and toxic at very low concentrations; and that simply counting the number of dead rats in an animal exposure is only the first step. It may turn out that sublethal effects—irritation and disorientation—are just as important. And we have learned that there are important delayed effects, that victims may appear to have survived only to be overwhelmed by lung disorders hours or days later.

This field is very active now and will be for some time to come. The development of toxicity guidelines for selection of materials for constructing and furnishing our living spaces has not yet begun. There are serious and difficult issues of regulation and enforcement, especially for furnishings.

Moveable furnishings in residences and in many public occupancies are not covered by codes nor any other control. Exceptions are carpets and mattresses covered by mandatory Federal standards under the Flammable Fabrics Act. The difficulties in controlling fire spread and growth in furnishings are two: the technology is not fully developed, and the means of implementing it are not tested.

Whereas ignition control may be effected for furnishings via the Consumer Product Safety Commission (CPSC), spread and growth of fire given a sustained ignition is not so readily achieved. This is because spread and growth involve interactions among several pieces of furnishings, each of which is purchased separately. No one may tell an owner of a house how many items made of a given material may be moved into a given room. The CPSC presumably has jurisdiction only over the individual items and not over their interactions. The building and fire codes do not apply except in special occupancies. To extend the coverage will require innovation. In addition to regulatory efforts, another possibility is educational labelling programs in which the potential contribution to fire is given. Such programs will take considerable ingenuity; nothing has been started as yet in this area because there remain technical problems to be solved.

Technically, the challenge is to develop a means of determining the fire severity in terms of escape time as a function of number, kind, and placement of furnishings and the geometry, interior finish, and ventilation in a room. This may be done in one of three ways: (a) full-scale room fire tests for each combination, (b) development of a series of small-scale tests shown to correlate rigorously to full-scale results, and (c) use of thermochemical and thermophysical properties combined with a mathematical model to predict escape time for a given configuration. Work is in progress on all three. We are sure of success on the first, but extensive requirements for full-scale testing would impose heavy financial burdens on industry. Nonetheless, the work is needed on full-scale room fire tests as the basis for the other two approaches. The use of a battery of small-scale tests run on each of the furnishings separately and the results integrated into one or more empirical predicting equations seems feasible in principle. The Center for Fire Research mounted a separate program just on this problem in late 1973. At that time, we estimated it to be a five-year project. We are now halfway through and are hopeful of success, though not yet sure of it. One of the difficulties is that many of the tests used in the past are not sufficiently precise, and some have given misleading results on synthetic polymers. It is our hope that we can upgrade some of these tests and remove the remaining difficulties by using the tests together rather than separately.

The more basic approach of using mathematical models developed from fluid mechanics is more satisfying but more difficult. The issue is not whether a model can be developed, but rather whether it will be sensitive enough to deal with the asymmetries and heterogeneities of the furnished room. There is no lack of ideas in this field, and we have at NBS a crack team on the problem. Time—perhaps five years—will tell.

Finally, we have the challenge of transferring to the one- and two-family residence the focus and concern for fire safety that has been devoted in the past to large buildings and public and special-risk occupancies. The fire community has addressed multiple occupancies because of the potential for major loss of life. When such disastrous fires occur, they command great public concern. On the other hand, most fire deaths occur in the home, one or a few at a time. If we are to cut our fire losses by half in the next generation, far more must be done to reduce residential fire fatalities. This means working through the local codes for the building itself and through CPSC and industry for the furnishings. It means innovations in design of new housing, better and less expensive detection and suppression systems, and far more effective programs to educate the public to the nature of fire hazards. The Department of Housing and Urban Development has a program to improve design concepts based on the National Fire Protection Association's (NFPA) decision tree approach. Our Center has a number of projects that bear on the question. The model codes and some local codes are pressing for wider use of detectors. But this is all a beginning. Vigorous and tenacious efforts will be required to raise the level of safety in the home to anywhere near that in most office buildings, industrial occupancies, and public places. Curiously, the fire safety in mobile homes may shortly be better than that in other residential housing simply because the fire problem in mobile homes is being addressed by the Department of Housing and Urban Development, as a result of new Federal law, as a single integrated problem. There may well be one or perhaps a series of mandatory standards issuing from this work.

These, then, are some of the current trends in fire research and the foundations on which they rest.

## DISCUSSION

*M. S. Abrams*[1] (*oral discussion*)—You showed a breakdown of the percentage of deaths in the various types of fires. The first line was residential at 72 percent. Do you have further breakdowns as to how much of that is single family, high rise, multifamily, low rise, public housing, and so forth?

*J. W. Lyons* (*author's oral closure*)—We do have available further breakdown as to the sort of fire that causes the problem in terms of occupancies. We don't have it in the manuscript. There is a paper coming out on this work, and the breakdown is there (Clarke, F.B. and Ottoson, John, *Fire Journal,* Vol. 70, [3], No. 20, May 1976). This is from the Fire Incident Data Organization system that NFPA operates. It is not the basis for their annual fire statistics. It is only on fire fatalities.

[1] Manager, Fire Research Section, Portland Cement Association, Skokie, Ill. 60076.

*Henry Tovey*[1]

# The National Fire Incident Reporting System: A Key to Fire Hazard Quantification

**REFERENCE:** Tovey, Henry, "**The National Fire Incident Reporting System: A Key to Fire Hazard Quantification,**" *Fire Standards and Safety, ASTM STP 614,* A. F. Robertson, Ed., American Society for Testing and Materials, 1977, pp. 11–25.

**ABSTRACT:** The National Fire Incident Reporting System is the collection of national fire loss data for fires attended by the fire service. It includes information on factors involved in fire ignition, spread, and extinguishment. It has been developed with recognition of the importance of serving the needs of state, regional, municipal, and local fire jurisdictions, as well as those at the Federal level. A representative selection of states has been identified for cooperative preliminary studies of this system. The system is based on use of the fire data reporting procedures developed by the National Fire Protection Association. Data elements have been selected as usual with many compromises to provide the most versatile use of data readily available. A review is presented of the way in which the data collected could be used in locating and analyzing fire problem areas as they occur. Brief mention is included on work under way to report and classify fire incidents on an international scale.

**KEY WORDS:** fires, fire hazards, information systems, mortality, fire losses

*America Burning,* the 1973 Report of the National Commission on Fire Prevention and Control, pointed up that, in the United States, fire is a major national problem [1].[2] It stated that fire claims nearly 12 000 lives a year, injures some 300 000, and destroys property worth over 4 billion dollars. It stated that the United States leads all the major industrialized countries in per capita deaths and property loss from fire. The death rate is nearly twice that of second-ranking Canada. Among the Commission's recommendations was that a National Fire Data System be established to help place fire prevention and control programs on a firmer foundation of scientific data and to facilitate development of cost-effective solutions to these problems.

[1] Director of Planning, National Fire Data Center, National Fire Prevention and Control Administration, U.S. Department of Commerce, Washington, D.C. 20236.
[2] The italic numbers in brackets refer to the list of references appended to this paper.

Acting on the recommendation of the Commission, the Congress passed and the President signed the Federal Fire Prevention and Control Act in 1974 [2]. The Act dealt with the entire national fire problem, and it established the National Fire Prevention and Control Administration (NFPCA). Federal-level responsibility for the collection, analysis, and dissemination of data on the occurrence, control, and results of fires of all types was assigned to the National Fire Data Center of the Administration. The legislative mandate also specified that the program of the Data Center shall be designed to provide an accurate nationwide analysis of the fire problem, identify major problem areas, assist in setting priorities, determine possible solutions to problems, and monitor the progress of programs to reduce fire losses. This paper describes the National Fire Incident Reporting System (NFIRS), an ongoing effort undertaken in partial fulfillment of this mandate, and it provides a very brief review of a related activity on the international level.

**The National Fire Incident Reporting System Concept and Operation**

The primary objective of the NFIRS is the collection of national fire loss statistics on fires attended by the fire service, including information on factors involved in fire ignition, spread, and extinguishment. However, the large cost of any system collecting national statistics led to an early policy decision that NFIRS must be more than a mere tool for data collection. Thus, the development or improvement of fire data systems serving the needs of state, regional, municipal, or local fire jurisdictions, as well as Federal needs, became a major NFIRS objective. Further considerations led to the establishment of another objective: the standardization and improved uniformity of fire data reporting and collection on all levels.

The NFIRS concept of a national fire data network is based on the cooperation of local, regional, and state fire jurisdictions, and the NFPCA (Fig. 1). Data for each fire incident are reported using a nationally uniform coding structure. The incident reports produced by the local fire service can be computer processed at the local or regional level and then passed on to the state jurisdiction in computerized form (Fig. 2). Alternatively, fire incident reports can be sent directly to the state jurisdiction, in which case they are computer processed by the state. All the fire data collected are tabulated and used by the state for problem analysis and production of annual and periodic reports, as well as feedback reports to the participating fire departments. On the state level, the computerized data are also processed into a national format and sent to the National Fire Data Center (Fig. 3). In the Center, the data received from all participating statewide systems are tabulated and analyzed, combined with other data bases, and reports are prepared for feedback to the state sources as well as dissemination to other interested groups.

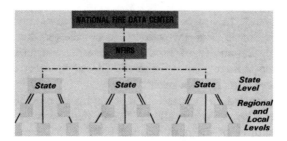

FIG. 1—*National Fire Incident Reporting System (an activity of the National Fire Data Center).*

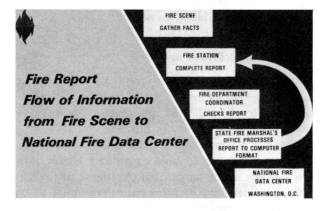

FIG. 2—*Fire report flow of information from fire scene to National Fire Data Center.*

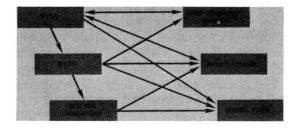

FIG. 3—*Flow of information (reports and analysis) from NFPCA.*

To ensure the maximum possible output versatility, the data coming to the Data Center are stored in the system on tapes essentially as they come in, without condensing. While some standard outputs for the system are already well defined, it is expected that the data base will be used by a broad group of users, with varied interests and points of view. In addition, important new uses for the data, not anticipated today, are still likely to be found. For this reason, it is believed that, at least initially, the system's capability of producing versatile outputs justify the somewhat higher storage costs. Since the initial number of data sources is limited, these costs are not prohibitive.

It is recognized, of course, that, in time, the NFIRS data base may become very large. When all the States that constitute a representative statistical sample of the nation join NFIRS, it will be possible to decrease the amount of data in storage. In addition, methods of summarizing and condensing much of the data will be developed in time, so that only a relatively small portion of the total data base will need to reside in costly storage. The remainder of all the data will be stored in an archival mode where it will be accessible when needed.

**Preliminary Test of the NFIRS Concept**

A preliminary test of the system concept was performed in 1974 and 1975 [3-5]. The test was limited to jurisdictions which already collected fire incident data. Results indicated that diverse fire service departments can prepare incident reports using a uniform classification scheme and that the system can accept reports from several different sources and produce a data base that can be used to produce useful outputs. During this period, it also became apparent that existing local fire service sources will be unable to provide fire incident data that would be representative and descriptive of the national fire experience, but that a number of states were interested in establishing statewide fire incident reporting systems for their own purposes. However, they lacked the necessary resources and needed assistance. These states indicated that they would be glad to cooperate with the NFPCA to ensure nationwide compatibility of the data for the benefit of their own state and the Nation. Accordingly, a decision was made to adopt officially a uniform classification scheme and a standard vocabulary; develop an official set of data elements and fire incident and fire casualty reporting forms; a training manual for instructors, and a handbook for those who will be filling out the forms; as well as a well-documented computer software package for processing the data. The computer package had to be complete, with instructions for keypunching the data from the report sheets at the front end and programs for producing reports meeting the basic requirements of the state and local level fire jurisdictions at the output end. It also had to be flexible to permit the different fire jurisdictions to modify it to fit their own special needs. Sample copies of the forms, the training manual/handbook, and the complete computer package are provided free of charge to state-level jurisdictions willing to participate. A set of states representative of the Nation was identified by a stratified statistical sampling plan, and these states have been assigned a high priority for participation in NFIRS.

## NFIRS Data Elements

Following the established policy guidelines, NFIRS has adopted the uniform classification scheme for fire data reporting developed by the Nation's fire community through the voluntary consensus mechanism of the National Fire Protection Association's Committee on Fire Reporting [6]. All data elements collected by NFIRS are based on this classification scheme, and all are included in the incident and casualty report forms developed by the Committee. The Committee also agreed to identify the data elements collected by NFIRS by small black triangles in the upper left-hand corners of the appropriate data element boxes. This permits fire jurisdictions that participate in NFIRS to use these forms readily, and the NFPCA adopted them (Figs. 4 and 5). It should be pointed up, however, that NFIRS does not require the use of these or any other specified forms. Each participating jurisdiction is free to design its own, as long as the data elements are based on the uniform classification scheme and the standard vocabulary adopted by NFIRS, and the form includes the nationally collected data elements.

While a sincere effort was made to make the selection of the data elements collected by NFIRS on the objective basis of utility and availability, inevitably the list had to be a compromise between those who wish to keep the incident report as simple and short as possible to minimize the effort necessary to fill it out and those who insist that comprehensive data on factors involved in fire ignition, spread, and extinguishment must be gathered to make possible the development of rationally justifiable and meaningful fire prevention and control programs. While only extensive experience will show how close the NFIRS set of data elements is to optimum, the following discussion indicates how the NFIRS elements relate to the objectives of the system in terms of their usefulness in identifying and characterizing the fire hazard.

Just what do we want to know about a fire when we try to identify and characterize the hazards associated with it? The answer to this question is not as easy as it may appear at first glance. However, while there may not be a consensus on this point, there appears to be an agreement on at least some of these elements: information on when and where the fire happened; on the causal factors; on fire growth; on the place in which the fire happened; on the total dollar loss; and on casualties. These categories will serve as examples.

The "when and where" information is provided by questions on Lines A and B of the NFPCA Incident Report Form. Information on date, day of week, and time of day makes it possible to correlate the frequency of

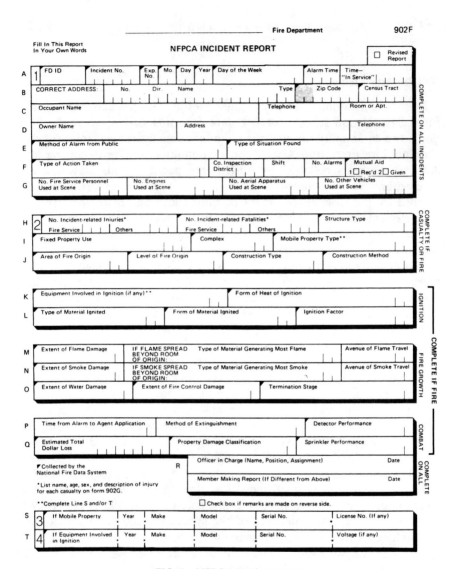

FIG. 4—*NFPCA incident report.*

incidents, their severity, and other characteristics, with temporal factors, such as daily activity patterns, or the seasons of the year. The zip code and Census Tract information provide the geographical coordinates which permit linkage of the fire data with, for example, demographic and socioeconomic data in the files of the Census Bureau. Information on causal factors is provided by the questions on Lines K and L of the form. The first of these is concerned with the ignition source and whether it

Fire Department

**NFPCA CASUALTY REPORT**

902G

Fill In This Report
In Your Own Words

| A | FD ID | Incident No. | Exp. No. | Mo. | Day | Year | Day of Week | Alarm Time | Page. . . . . . of . . . . . . . |

**CASUALTY 1**

| | Casualty Number | ☐ Revised Report |

| U | Casualty Last Name | First Name | MI | D.O.B. | Age | Time of Injury |
| V | Home Address | | | Telephone | |

| W | SEX | CASUALTY TYPE | SEVERITY | AFFILIATION |
| | 1☐ Male 2☐ Female | 1☐ Fire Casualty 2☐ Action Casualty 3☐ EMS Casualty | 1☐ Injury 2☐ Death | 1☐ Fire Service 2☐ Other Emergency Personnel 3☐ Civilian |

| X | Familiarity With Structure | Location at Ignition | Condition Before Injury |
| Y | Conditions Preventing Escape | Activity at Time of Injury | Cause of Injury |
| Z | Nature of Injury | Part of Body Injured | Disposition |

☐ See Remarks on Back          ☐ See Additional Report

**CASUALTY 2**

| | Casualty Number | ☐ Revised Report |

| U | Casualty Last Name | First Name | MI | D.O.B. | Age | Time of Injury |
| V | Home Address | | | Telephone | |

| W | SEX | CASUALTY TYPE | SEVERITY | AFFILIATION |
| | 1☐ Male 2☐ Female | 1☐ Fire Casualty 2☐ Action Casualty 3☐ EMS Casualty | 1☐ Injury 2☐ Death | 1☐ Fire Service 2☐ Other Emergency Personnel 3☐ Civilian |

| X | Familiarity With Structure | Location at Ignition | Condition Before Injury |
| Y | Conditions Preventing Escape | Activity at Time of Injury | Cause of Injury |
| Z | Nature of Injury | Part of Body Injured | Disposition |

☐ See Remarks on Back          ☐ See Additional Report

**CASUALTY 3**

| | Casualty Number | ☐ Revised Report |

| U | Casualty Last Name | First Name | MI | D.O.B. | Age | Time of Injury |
| V | Home Address | | | Telephone | |

| W | SEX | CASUALTY TYPE | SEVERITY | AFFILIATION |
| | 1☐ Male 2☐ Female | 1☐ Fire Casualty 2☐ Action Casualty 3☐ EMS Casualty | 1☐ Injury 2☐ Death | 1☐ Fire Service 2☐ Other Emergency Personnel 3☐ Civilian |

| X | Familiarity With Structure | Location at Ignition | Condition Before Injury |
| Y | Conditions Preventing Escape | Activity at Time of Injury | Cause of Injury |
| Z | Nature of Injury | Part of Body Injured | Disposition |

☐ See Remarks on Back          ☐ See Additional Report

| R | Officer in Charge (Name, Position, Assignment) | Date |
| | Member Making Report (If Different From Above) | Date |

☛ Collected by the
National Fire Data System

FIG. 5—*NFPCA casualty report.*

was a piece of equipment. Information on this point is of vital importance to equipment manufacturers, as well as to regulatory agencies such as the Consumer Product Safety Commission, because it identifies equipment and products that require further attention. For this reason, if it is a piece of equipment that was involved in ignition, further questions are asked about it on Line T of the form (the make, year, model, serial number, and voltage).

The second question in this series asks about the form of the ignition energy, whether it was the heat from a burning cigarette, or a spark from an open fire. The next two questions are concerned with the nature and the form of the first item ignited, and the final one is concerned with the condition or situation—the action or lack of action—that permitted the heat from the ignition source to cause the ignition of the first item.

All these questions have been very carefully defined and selected, so that, when the answers are considered together, they can provide the information necessary for reconstructing the causal chain of events that led to the fire. For example, if the equipment involved was an "electric iron," the heat of ignition was "heat from properly operating electrical equipment," the type of material first ignited was "wearing apparel not on a person," and the ignition factor was "unattended operation," the cause of fire can be readily visualized.

Information on fire growth is provided by the questions on Lines M, N, and O. It is made up of several components, and, like the information on fire cause, it is most useful when these components are related to each other and to other information about the fire. One component is concerned with the extent of the spread: was the flame, or smoke, or water damage confined to the object of origin? to the room of origin? to the floor? Did it extent beyond the building of origin? Obviously, this information in itself has a limited value. But, if a high correlation were found between extent of smoke damage and, say, polystyrene as the type of material first ignited, this would have some interesting implications concerning the fire hazard of polystyrene products.

The second component of the fire growth complex is concerned with what helped it most to grow. After all, it is not always the first item ignited that is the culprit, the important factor in fire spread. It could have been the second or even the third item ignited, the drapery that spread the fire across the room in no time at all, but was ignited from a small wastebasket fire started by a glowing match. The third component is concerned with the path taken by the spreading fire: did it move up an open staircase? Did it spread across a long hall without fire doors? Or did it move through to a hole in a fire wall? The collecting of information on these two aspects of fire growth is a relatively new development in fire reporting, and, during the initial stages, the system will not accumulate it on the national level. However, because the inferences that can be drawn from data on these points should be of much value to those concerned with life safety codes, architects, builders, makers of furniture, furnishings, and countless others, the system provides for recording and collection of the data on the state level.

Information on the place where the fire happened is provided by a series of questions on Lines H, I, and J. They deal with type of structure in which the fire occurred, the construction type and method, and the use to which the property was put at the time of the fire. Data derived from answers to the first of these questions permits the analysis of fire incidence statistics in terms of the type of structures involved, such as buildings, tents, bridges, or underground structures. Information on construction type permits analysis in terms of standard constructions that differ in fire resistance and stability under fire conditions, such as fire resistive structures, heavy timber, or unprotected wood frame. The terminology used in the system is based on a National Fire Protection Association (NFPA) standard [7], which can be related to the several building codes in use in the United States, such as the Basic Building Code, Standard Building Code, and the Uniform Building Code. The data should, therefore, be useful for establishing code requirements and monitoring their effectiveness.

Data on methods of construction are meant to help in identifying differences that may exist between the behavior in fire of structures built on the site and those built in a factory in a modular form or assembled on the site.

Information on fixed property use allows analysis of the fire problem on the basis of property use, one of the most popular ways of analyzing fire data. Many of the codes and regulations and much of the effort in the fire area are directed at problems presented in terms of a particular type of occupancy such as manufacturing plants, restaurants, nursing homes, hospitals, apartments, or single-family dwellings. The rational way of establishing the criteria for these codes and regulations and of allocating priorities to these efforts is based on the fire hazard associated with each occupancy type as identified and quantified by fire experience data.

Information on the dollar losses resulting from fires is one major factor in determining priorities for, and determining the cost-effectiveness of, programs aimed at combatting the fire problem. Unfortunately, reliable dollar loss data are hard to come by because fire service personnel are not schooled appraisers. The multitude of methods for loss estimates—original cost, market value, replacement cost—is another obstacle to obtaining comparable data. Since information on total dollar loss is so important, both a dollar estimate and a range are asked for on Line Q of the form. The instructions explicitly request that the estimates be made on the basis of cost of replacement in like kind and quality and that only the direct physical loss to the structure, contents, machinery, equipment, and such, be considered.

The identification and characterization of casualties resulting from a fire have always been integral parts of fire reporting. NFIRS collects information on fire casualties on a separate casualty report (Fig. 5) which calls for information on the victim's age, sex, affiliation, and casualty type and severity. Other questions, such as those concerning the nature of the injury, part of body injured, and disposition, or those concerning familiarity with structure and conditions preventing escape, are designed to provide additional information about fire casualties for analysis and correlation with the various fire incident parameters. For example, if the data show that lack of familiarity with the structure is an important factor, an educational campaign to get people familiar with the structures where they live and work in order to cut down the number of casualties would seem indicated. However, if correlation with incident data shows that lack of familiarity with structure is of importance only for certain types of occupancies, such campaigns could be aimed more precisely and thus be both more effective and more cost effective.

**National Fire Data System**

This discussion did not cover all the data elements on the NFPCA incident reporting form, but it should be apparent from it that, when NFIRS is implemented on a scale sufficient to provide data statistically representative of the national fire experience, and the data quality is good, it will be possible to identify the major fire hazards and rank them in order of priority. Still, it should be made clear that, while NFIRS is expected to fulfill a significant part of NFPCA's mandate in the fire data area, it was not designed to satisfy all fire data needs. The National Fire Data Center is developing several other data systems which will supplement NFIRS and together constitute a comprehensive National Fire Data System. Thus, since NFIRS is limited by design to fire incidents attended by the fire service, household surveys, similar to that described in a recent NFPCA publication [8], will be conducted periodically. Also, since NFIRS was not designed to provide the detailed, exhaustive information necessary for suggesting possible solutions to a fire hazard problem, the Data Center has planned and is implementing a network for in-depth investigation of specified classes of fires, conducted on a contract basis by well-trained investigators. This effort will be coordinated with a related program conducted at the Center for Fire Research, National Bureau of Standards.

The National Fire Data System is expected to include relevant fire data from the insurance industry, National Center for Health Statistics, the Consumer Product Safety Commission, and other governmental and private sources. In addition, it will incorporate or establish regular or direct access to other data files necessary for a full understanding of the causes and effects of fires, such as the demographic and socioeconomic files of the Census Bureau.

**International Fire Data Network**

It is also hoped that, at some time in the future, the U.S. National Fire Data System will be a part of an international network of fire data systems. Fire is an international problem, but, at the present time, it is extremely difficult to compare fire losses in one country with fire losses in another. The statistics, if collected at all, are collected on different bases, with each country including or excluding different types of fire, using its own definitions, and classifying the data in different ways. Reliable statistics on fires involving specific products, in particular, are very scarce. However, efforts to develop a basic international system for fire data, capable of producing statistics that could be compared from country to country, are underway under the auspices of the International Standards Organization. Because textile flammability has been recognized as a problem for some years now, the Subcommittee on Burning Behavior of Textiles and Textile Products of the Committee on Textiles assumed responsibility for the effort, with the actual work performed by the Subcommittee's Working Group 6, on Risk Data Analysis (ISO/TC38/SC19/WG 6). The Working Group has held three meetings over the past couple of years: in Zurich in June 1974 and in Paris in October 1974 and May 1975. The meetings were attended by experts from Australia, Belgium, Canada, France, Germany, Italy, Japan, Netherlands, Norway, Switzerland, the United Kingdom, and the United States. The meetings of the parent subcommittee were also attended by representatives of Austria, Denmark, Finland, South Africa, Spain, and Sweden, so that these countries also had an opportunity to participate in the effort.

We have exchanged information about the status of fire data systems in the various countries, and I might mention that they range all the way from rudimentary ones to sophisticated systems such as that in France, which is capable of producing complex reports and answering queries.

Having become acquainted with one another and with fire data systems in each other's countries, the members of the Working Group agreed that, because it would be impractical to collect detailed data on all fire incidents, it is necessary to establish a bilevel system. The concept is similar to that adopted in the United States in that the system would collect statistical data on all fires and detailed data through in-depth investigation of an appropriate sample of fires of a specified type. The Working Group also agreed that maximum use should be made of existing systems and other existing resources.

To achieve these objectives, the Working Group agreed that its members will send information on statistical data on all fires to the United States—to the National Fire Data Center—for a preliminary examination and perhaps the development of a tentative list of data elements to be used in a standard international fire incident reporting form and send detailed reports on textile reports to Dr. Janet Thompson of the United

Kingdom, for the same purpose. Following the Working Group's meetings, the Subcommittee reviewed these plans in a plenary session and formally endorsed the effort to examine data from a number of different countries, evaluate it, and try to propose standardized international forms for reporting fire incidents.

The exact date and place of the next meeting of WG-6 has not yet been set, although the U.S. delegation has issued an official invitation to the Subcommittee to meet in the United States in September 1977. This is more than a year from now, but the tasks undertaken by the subcommittee and its Risk Data Analysis Working Group are very substantial, and we are already behind schedule. Still, progress is being made. We have been able to define what we want to do and agreed on how to do it. This, I am told, is half the battle.

I am optimistic that it will not be very long before standard national and international fire statistics become available and permit us to learn what is the magnitude of the fire problem, what are the high-priority areas for standards, codes, public education, and research. Through data analysis, we will be able to contribute to the solving of the nation's and the world's fire problems.

## References

[1] *America Burning,* report of the Commission on Fire Prevention and Control, 1973. Superintendent of Documents, U.S. Government Printing Office, Washington, D.C. 20402.
[2] The Federal Fire Prevention and Control Act of 1974, PL 93-498.
[3] Tovey, Henry, *Fire Journal,* Nov. 1974, pp. 91–96.
[4] Eisenberg, D. E. and Getis, R. L., "Designing to Capitalize on Existing Resources," Report 7500-522, Auerbach Associates, Inc., 1974.
[5] Buchbinder, Benjamin, *Fire Journal,* May 1975, pp. 65–69.
[6] Report on Committee on Fire Reporting, NFPA Technical Committee Reports, National Fire Protection Association, 1975, pp. 415–514.
[7] Report of the NFPA Committee on Building Construction, adopted at the 5 May 1975 Annual Meeting of the National Fire Protection Association, and published in Ref 6. It is scheduled for publication in the 1976 edition of the *National Fire Codes.*
[8] Highlights of the National Household Fire Survey, The National Fire Prevention and Control Administration, 1975.

# DISCUSSION

*K. P. Reynolds*[1] (*oral discussion*)—When will this be a national program where all fire departments will feed information into the program?

*Henry Tovey* (*author's oral closure*)—Well, the time schedule is somewhat like this. We are now working with three states in the validation

[1] County fire marshal, Albemarle County, Charlottesville, Va. 22901.

program. We hope to complete this in three months, and, by the end of the fiscal year, we hope to add another four to five states. We hope to get eight additional states within the next year. A major factor that must be accounted for in this effort is the readiness of a state to install and operate a state-wide fire incident reporting system. Incidentally, we have had several inquiries about NFIRS from various fire chiefs in Virginia, and my answer is always: please get together. Unfortunately, we are in no position to deal with individual fire departments. We simply do not have the resources and staff that would be required.

*R. P. Fleming*[2] (*oral discussion*)—I didn't notice a space on the Fire Incident Report where one could indicate whether the point of fire origin was protected by an automatic fire suppression system. Do you not think that this information is crucial to any analysis of fire growth in regard to constructing type?

*Henry Tovey* (*author's oral closure*)—The answer is yes, there is a place (see Line Q sprinkler performance), and the answer is yes, we do feel it is crucial. Now, how frequently will we get that information? I don't know, but we will find out in a little while.

*C. R. Williams*[3] (*oral discussion*)—When might the first results from your study be available?

*Henry Tovey* (*author's oral closure*)—We hope to have the first reports out by the end of this calendar year. They won't be very good.

*M. Lanier*[4] (*written discussion*)—There is no doubt that NFIRS is a needed data-gathering program; however, many states and especially local governments do not have funds available to establish and maintain such a program. Because of this, it seems that states and local governments will need Federal support to assist in securing people or in training people to set up and maintain the system.

Georgia is a highly rural state with an estimated 450 fire departments. An estimated 75 to 80 percent of these are small-town volunteer departments. Most have no records or, at best, very poor records. Due to this type of situation, the establishment of a "statewide" reporting system would be a costly and time-consuming matter.

Is it intended by the NFPCA for "all" fire departments and states to become a part of a fire reporting system? If not, what percentage of departments from cities and what percentage of departments from rural areas would be acceptable to permit realistic estimates of the fire problem. Are Federal funds available to states and local governments to set up a NFIRS?

*Henry Tovey* (*author's written closure*)—The National Fire Prevention

[2] Director, Engineering Standards, National Automatic Sprinkler and Fire Control Association, Mt. Kisco, N.Y. 10549.

[3] Commercial development manager, Monsanto, St. Louis, Mo. 63166.

[4] Fire marshal, Rome Fire Department, Rome, Ga. 30161.

and Control Administration recognizes that many states and local governments do not have the funds required to establish and maintain statewide fire incident reporting systems. For this reason, the NFPCA established a State Fire Incident Reporting Assistance Program. Under this program, the NFPCA provides, upon the approval of a grant request from a state, master copies of forms suitable for use in fire incident and casualty data collection; master copies of training materials; a computer software package for processing the data; and financial assistance ranging from $5000 to $20 000 per state, depending on need. Admittedly, the funds provided under this program are not nearly adequate to offset the cost incurred by a state when it establishes a statewide fire incident reporting system. However, the NFPCA must also operate within a very limited budget. A detailed description of the state fire incident reporting assistance program, with guidelines for applying for a grant, can be obtained by writing to the NFPCA, P. O. Box 19518, Washington, D.C. 20036.

The other question was whether NFPCA intends all fire departments to become participants in the National Fire Incident Reporting System. This is a worthy objective, but we fully recognize that it is not likely to be achieved in the near future. For this reason, NFPCA has identified 19 states which, when taken together, constitute a satistically representative sample in the United States with regard to fire experience. These states include California, Colorado, Delaware, Florida, Iowa, Louisiana, Maryland, Massachusetts, Michigan, Minnesota, Missouri, Nevada, New York, Ohio, Oregon, Rhode Island, South Dakota, Washington, and West Virginia. Because of its limited resources, NFPCA assigns a higher priority to these states than to states not on the list. Since some of the states on the list are not now and are not likely to be in a position to go statewide in the near future, it may well be necessary to resort to statistical sampling for obtaining representative data from these states. Such methods have not yet been developed.

*J. Deitz*[5] (*written discussion*)—There are two groups, which, from the discussion, do not appear to be included in the data base collection system. These include the Federal Government (military bases, government installations with their own fire protection facilities) and private industry (large complexes where most fires are not reported to a municipal authority). Is any effort being made to include these in the data base?

*Henry Tovey* (*author's written closure*)—Your remark that the National Fire Incident Reporting System is limited to data on fires attended by state and local government fire service is quite correct. The National Fire Data Center is planning to supplement these data with data obtained from other sources. To respond specifically to your questions, data on

[5] Fire protection engineer, Brookhaven National Laboratory, Upton, N.Y. 11973.

fire experience in government installations are being collected through the mechanism of the former Federal Fire Council; and we are attempting to collect data on the fire experience in large industrial installations, protected by private fire service. These attempts are still in the early stages, and any assistance anyone can provide in this effort would be more than welcome. Our hope is that, ultimately, all Federal agencies as well as private fire departments will use fire incident reporting forms which contain the data elements on the NFIRS form and will provide the data to the Data Center.

*W. G. Berl*[1] *and B. M. Halpin*[1]

# Fire-Related Fatalities: An Analysis of Their Demography, Physical Origins, and Medical Causes

---

**REFERENCE:** Berl, W. G. and Halpin, B. M., "**Fire-Related Fatalities: An Analysis of Their Demography, Physical Origins, and Medical Causes,**" *Fire Standards and Safety, ASTM STP 614*, A. F. Robertson, Ed., American Society for Testing and Materials, 1977, pp. 26–54.

**ABSTRACT:** Data are presented and analyzed on the demographic variables (location, age, sex, and race) of people perishing in fires in the state of Maryland in 1972/1973, on the physical causes of these fires, and on the most likely medical effects. Carbon monoxide as asphyxiant, cigarettes as ignition source, and the disproportionately large involvement of children and the elderly are pointed up. The contributory effects of alcohol ingestion, particularly in men in the 40 to 60 years age group, are described.

**KEY WORDS:** fires, demographic surveys, mortality, fire hazards, toxicity

Many of the materials used in buildings, such as brick, stone, concrete, glass, or steel, will not sustain combustion. But a typical dwelling contains many combustible materials, such as wood, plastics, paints, paper, textiles, and metals as part of its structure and furnishings and so many energy sources that can ignite them, that it is not surprising that accidental and unwanted fires are a common occurrence. Whether such incidents remain innocuous or develop into fires causing substantial damage depends on intricate combinations of fuel types and geometry, complex and mostly unpredictable interactions of heat and gas flows, and on many other parameters. It also depends on the effectiveness of the countermeasures that are called into action.

The number of incidents requiring concerted countermeasures is great. A large body of statistical information is, in principle, available, the analysis of which could lead to insights into the most common causes of

---

[1] Principal investigator, Fire Problems Program and systems analyst, respectively, Applied Physics Laboratory, The Johns Hopkins University, Silver Spring, Md. 20910.

fires, their rates of spread, the behavior of humans under stress, and the economic and human losses.[2] Unfortunately, the full power of this informational resource has rarely been exploited. As a consequence, productive recommendations for better building and products designs, for improved countermeasures, or for safer human behavior patterns have yet to come forth.

Of all the unwanted fires, those that result in death to human beings are subject to the most intensive investigations into their origins and consequences. Their numbers are sufficiently large (that is, several thousand per year) to provide a sound statistical base without raising the data gathering and interpreting to an unmanageable size.[3] They represent the extreme outcomes of a much larger and even more diverse number of urban fires with less severe consequences but with similar case histories. However, in this area, detailed and extended analyses that tie together causes, consequences, and demographic information are also very rare.

The ongoing investigation of fire-related fatalities, summarized in this paper and carried on since the fall of 1971, has as its objective the obtainment of insights into the demographic characteristics of fire victims, the most probable physical and psychological causes of fire exposures, and the biochemical reasons for fire deaths. The results that we wish to obtain from this study are whether and in what ways substantial reductions in the number of fatalities can be effected. It also provides a data base against which to compare future trends.

The study is limited to fatalities that have occurred in the state of Maryland during 1972 and 1973. This restriction was set by the fact that this geographical area provides an adequately large number of cases (approximately 100/year) and because the necessary medical and legal permissions for carrying out the study could be obtained satisfactorily within this political unit. It was further restricted to casualties that occurred

---

[2] A recent *National Household Fire Survey* (highlights published by the U.S. Department of Commerce/National Fire Prevention and Control Administration) gives an estimate of 5 575 000 unintentional fires/year in the United States, based on a sampling of 30 500 households. An incident was counted as a fire if it emitted smoke or flames and was not started intentionally. Less than 10 percent of the fires required the attention of fire departments. The most detailed and consistent analyses of the causes and extinction methods of fires reported to fire departments are published in the *Annual Reports of the Director of Fire Research,* Department of the Environment and Fire Officers' Committee, Joint Fire Research Organization, London.

[3] Fire deaths in the United States are reported annually in the September issues of *Fire Journal* (published by the National Fire Protection Association). Multiple-death fires, accounting for approximately 10 percent of all fatalities, are published in the July issues. The total fire deaths (including motor vehicle fires) are given as 5.71/100 000 for 1972. A critique of the shortcomings of available fire fatality data is given in Halpin, B. M., "Survey and Evaluation of U.S. National Fire Death Fatalities," APL/JHU FPPTR6, The Johns Hopkins University/Applied Physics Laboratory, Dec. 1971.

within 6 h of the onset of the fire, thus excluding fatalities that occurred at a later time and which are covered in a separate investigation.[4]

In view of the paucity of detailed fire statistics on fire injuries and deaths, we are unable to compare the findings of this study with results from other sections of the United States or from other countries so as to establish similarities or differences with experiences elsewhere. However, we believe Maryland to be a typical cross section of the United States in terms of the social and economic patterns that give rise to fire accidents and conclude that the observed results probably have general validity. A report on much of the information discussed here has been published in Footnote 5.

**Data Acquisition**

The data used in this paper were obtained from several sources. Demographic data came from reports of fire department and fire marshal investigators who provided information about the location of the fires and the personal data of the fire victims. The most probable causes of the fire starts were also obtained from the assessments of the investigators. Biochemical and clinical information regarding the medical causes of death were derived from detailed autopsies of the fire victims. Additional information about materials first ignited or involved in the fire spread, escape attempts, ingestion of fire gases, soots, and of heavy metal oxides were also available.

For analysis, the data were subdivided into three distinct geographical areas: (a) urban Baltimore city, (b) the four large counties (Baltimore, Prince Georges, Montgomery, and Anne Arundel) which are predominantly suburban but with some urban and rural characteristics, and (c) the 18 small counties which are mostly rural. The population distribution, the number of fire fatalities, and the number of autopsies performed in these three areas in 1972 and 1973 are given in Table 1.

The biochemical and clinical follow-up was most detailed for fire victims in Baltimore city where 59.7 percent of the fatalities met the 6-h time and the autopsy criteria (Table 1). In order to avoid a predominance of city-oriented information, the data were normalized (that is, reduced for Baltimore city and increased for the 18 small counties) so as to match, on a percentage basis, the total reported fire casualties in the various administrative subdivisions within the state of Maryland. The

[4]Preliminary information (Annual Summary Report 1 July 1973–30 June 1974, APL/JHU FPPA74, The Johns Hopkins University/Applied Physics Laboratory, Aug. 1974) indicates that the ratio of fire deaths that occur in Maryland later than 6 h after the onset of fire to those occurring in less than 6 h is approximately 1/2.5. This results in a 71 percent coverage of fire deaths in this study.

[5]Halpin, B. M., Radford, E. P., Fisher, R., and Caplan, Y., "A Fire Fatality Study," *Fire Journal*, Vol. 69, 1975, p. 11.

TABLE 1—*Population and fire fatality data in Maryland (1972/1973).*

| Location | Population | Total Fire Fatalities[a] (two-year period) | Annual Fatalities Per 10 000 Population | Total Autopsies Performed[b] | Fatalities Autopsied, % |
|---|---|---|---|---|---|
| Maryland | 3 910 000 | 257 | 3.28 | 100 | 38.9 |
| Baltimore | 906 000 | 72 | 3.95 | 43 | 59.7 |
| 4 large counties | 2 104 000 | 110 | 2.60 | 38 | 34.6 |
| 18 small counties | 900 000 | 75 | 4.16 | 19 | 25.3 |

[a] As reported by the Maryland Fire Marshal.
[b] Carried out by the Medical Examiner's Office, State of Maryland.

data were correlated in a number of different ways. The "analysis tree" followed in this paper is indicated in Fig. 1.

**Per Capita Fatalities**

The overall fire fatalities statistics in 1972/1973 for the state of Maryland and of its 23 administrative subdivisions (Table 2) lead to the following conclusions.

The average fire fatalities for the state were 3.28/100 000/year. The per capita casualties were highest (4.16/100 000/year) in the 18 small counties, followed closely by the densely populated Baltimore city (3.95). The lowest per capita fatalities (2.6) were found in the four large counties. The reported U.S. fatality rate is 5.71/100 000/year.[3]

While the average per capita fatalities in rural counties were high, wide fluctuations exist among them (Table 2). Two thirds of them fall above and one third below the mean for the state of Maryland. These fluctuations are, in part, due to the sporadic occurrence of fire fatalities. However, the particularly high and constant fatality record of Cecil County (more than double the state average) indicates a specific problem area. If, by appropriate changes, the life safety record in the state of Maryland could be lowered to match the current best record of the suburban counties, a 21 percent reduction in the annual number of fatalities would occur. Larger reductions would require additional improvements in the fatality record in every geographical area.

**Demographic Factors**

The absolute numbers of fire fatalities by location, age, sex, and race are given in Table 3. In Figs. 2*a,b* (age and location), Figs. 3*a,* 3*b* (sex, race, and age), and Fig. 4 (sex, race, and location), the data, adjusted for the autopsy variance in the three geographical areas, are compared.

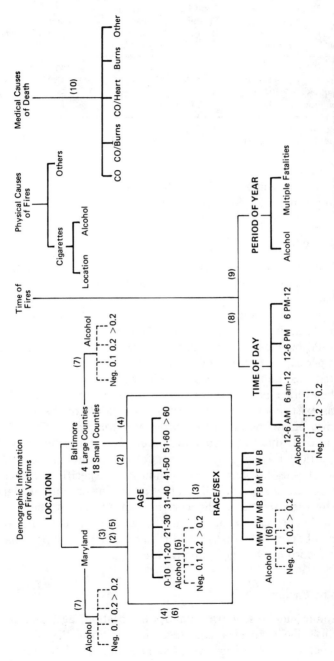

FIG. 1—"Analysis tree" of fire fatalities. Numbers in parenthesis refer to figures in the text in which specific information is displayed.

TABLE 2—*Fatalities in political subdivisions of the state of Maryland.*

| Subdivision | Population × 10³ | Fatalities Two-Year Period | (1972/1973) | Annual Fatalities Per 100 000 Population |
|---|---|---|---|---|
| Caroline | 11 | 5 | (0/5) | 22.7 |
| Cecil | 53 | 11 | (6/5) | 10.4 |
| Somerset | 19 | 4 | (2/2) | 10.5 |
| Queen Anne | 18 | 3 | (0/3) | 8.3 |
| Worcester | 24 | 4 | (4/0) | 8.3 |
| Alleghany | 84 | 11 | (9/2) | 6.5 |
| St. Mary | 47 | 5 | (3/2) | 5.3 |
| Dorchester | 29 | 3 | (2/1) | 5.2 |
| Harford | 115 | 10 | (6/4) | 4.3 |
| Charles | 48 | 4 | (1/3) | 4.2 |
| Baltimore City | 906 | 72 | (39/33) | 4.0 |
| Washington | 103 | 7 | (1/6) | 3.4 |
| Prince Georges | 661 | 43 | (21/22) | 3.3 |
| Anne Arundel | 298 | 18 | (10/8) | 3.0 |
| Talbert | 23 | 1 | (0/1) | 2.2 |
| Baltimore County | 622 | 27 | (15/12) | 2.2 |
| Montgomery | 523 | 22 | (12/10) | 2.1 |
| Wicomico | 54 | 2 | (2/0) | 1.9 |
| Frederick | 84 | 2 | (2/0) | 1.2 |
| Carroll | 69 | 0 | (0/0) | 0 |
| Garrett | 21 | 0 | (0/0) | 0 |
| Howard | 62 | 0 | (0/0) | 0 |
| Kent | 16 | 0 | (0/0) | 0 |
| | | 254 | (135/119) | |

## Age

The percentage of fatalities in the age groups 0 to 10, 51 to 60, and above 60 are substantially higher nearly everywhere in the State than what would be predicted on the basis of the U.S. Census population-age distribution (Figs. 2*a,b,* and Table 4). The age distribution in Maryland, according to 1970 Census data, is given in Table 4. The children fatalities in Baltimore city and fatalities among people older than 65 in the 18 small counties are particularly high. For example, in Baltimore city, 27.7 percent of all fatalities occur in children under 10 years of age, whereas only 18 percent are expected, based on their number in the population. In the age group 0 to 2 years, the fatalities record is substantially poorer. In contrast, fatalities in the age group 11 to 20 (9.2 percent) is markedly below the level expected from population distribution. It is reasonable to assume that the high children fatalities are influenced by their inability to escape a fire on their own (especially below the age of two), their lack of judgment about escape routes, and the large amount of time spent by them indoors. The same arguments probably apply to the age

TABLE 3—Autopsied fire fatalities (100 cases) by location, age, sex, and race (1972/1973).

| Age, years | Male | Female | White | Black | Male White | Female White | Male Black | Female Black | Sum |
|---|---|---|---|---|---|---|---|---|---|
| *Maryland* | | | | | | | | | |
| 0 to 10 | 11 | 13 | 10 | 14 | 4 | 6 | 7 | 7 | 24 |
| 11 to 20 | 7 | 4 | 10 | 1 | 6 | 4 | 1 | 0 | 11 |
| 21 to 30 | 5 | 7 | 8 | 4 | 4 | 4 | 1 | 3 | 12 |
| 31 to 40 | 6 | 2 | 6 | 2 | 5 | 1 | 4 | 1 | 8 |
| 41 to 50 | 9 | 3 | 8 | 4 | 5 | 3 | 4 | 0 | 12 |
| 51 to 60 | 11 | 6 | 13 | 4 | 9 | 4 | 2 | 2 | 17 |
| >60 | 8 | 8 | 12 | 4 | 5 | 7 | 3 | 1 | 16 |
| Total | 57 | 43 | 67 | 33 | 38 | 29 | 19 | 14 | 100 |
| *Baltimore* | | | | | | | | | |
| 0 to 10 | 7 | 5 | 2 | 10 | 2 | 0 | 5 | 5 | 12 |
| 11 to 20 | 3 | 1 | 3 | 1 | 2 | 1 | 1 | 0 | 4 |
| 21 to 30 | 3 | 2 | 2 | 3 | 2 | 0 | 1 | 2 | 5 |
| 31 to 40 | 5 | 1 | 5 | 1 | 4 | 1 | 1 | 0 | 6 |
| 41 to 50 | 4 | 1 | 2 | 3 | 1 | 1 | 3 | 0 | 5 |
| 51 to 60 | 4 | 2 | 4 | 2 | 4 | 0 | 0 | 2 | 6 |
| >60 | 2 | 7 | 3 | 2 | 1 | 2 | 1 | 1 | 5 |
| Total | 28 | 15 | 21 | 22 | 16 | 5 | 12 | 10 | 43 |
| *4 Large Counties* | | | | | | | | | |
| 0 to 10 | 3 | 6 | 7 | 2 | 2 | 5 | 1 | 1 | 9 |
| 11 to 20 | 2 | 2 | 4 | 0 | 2 | 7 | 0 | 0 | 4 |
| 21 to 30 | 2 | 4 | 5 | 1 | 2 | 3 | 0 | 1 | 6 |
| 31 to 40 | 0 | 1 | 0 | 1 | 0 | 0 | 0 | 1 | 1 |
| 41 to 50 | 2 | 2 | 4 | 0 | 2 | 2 | 0 | 0 | 4 |
| 51 to 60 | 6 | 2 | 7 | 1 | 5 | 2 | 1 | 0 | 8 |
| >60 | 4 | 2 | 5 | 1 | 3 | 2 | 1 | 0 | 6 |
| Total | 19 | 19 | 32 | 6 | 16 | 16 | 3 | 3 | 38 |

*18 Small Counties*

| | | | | | | | | | |
|---|---|---|---|---|---|---|---|---|---|
| 0 to 10 | 1 | 2 | 1 | 2 | 0 | 1 | 1 | 1 | 3 |
| 11 to 20 | 2 | 1 | 3 | 0 | 2 | 1 | 0 | 0 | 3 |
| 21 t0 30 | 0 | 1 | 1 | 0 | 0 | 1 | 0 | 0 | 1 |
| 31 to 40 | 1 | 0 | 1 | 0 | 1 | 0 | 0 | 0 | 0 |
| 41 to 50 | 3 | 0 | 2 | 1 | 2 | 0 | 1 | 0 | 3 |
| 51 to 60 | 1 | 2 | 2 | 1 | 0 | 2 | 1 | 0 | 3 |
| >60 | 2 | 3 | 4 | 1 | 1 | 3 | 1 | 0 | 5 |
| Total | 10 | 9 | 14 | 5 | 6 | 8 | 4 | 1 | 19 |

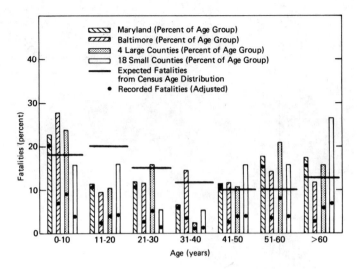

FIG. 2a—*Correlations of fire fatalities as function of age.*

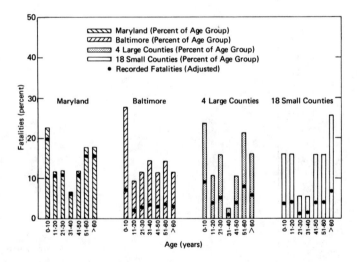

FIG. 2b—*Correlations of fire fatalities as function of location.*

group above 65 years. On the other hand, the very high risk shown by the 51 to 60 age group is associated with a particularly high incidence of alcohol intoxication (see later) which is most likely a contributory factor both to the large number of ignitions (from cigarettes) and to the difficulty of fashioning proper responses for escape.

## Sex

Analysis of the data related to the sex of fire victims indicates the

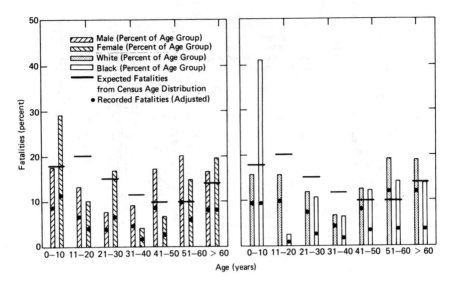

FIG. 3a—*Correlations of fire fatalities as function of age.*

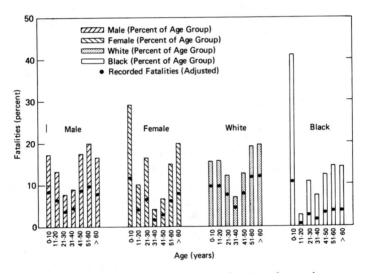

FIG. 3b—*Correlations of fire fatalities as function of sex and race.*

following. Males present a greater risk at all ages (Figs. 3a,b) and locations (Fig. 4) than females (57/43), except for a high female child fatality
rate that may be due to greater flammability of dresses worn by young
girls, and a female preponderance in the 21 to 30 age group. As will be
shown later, alcohol consumption, largely by men, begins around the age
30 and becomes increasingly pronounced up to the age of 60. This heavy

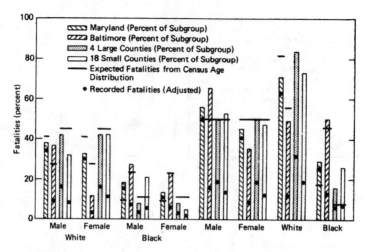

FIG. 4—*Correlation of fire fatalities in Maryland as function of sex and race* (*M = male, F = female, W = white, B = black*).

drinking habit of older men not only contributes to the increase in the absolute number of fatalities in the older age groups, but also shifts the male/female ratio markedly toward men. It is a behavior pattern that is particularly prevalent in Baltimore city.

*Race*

With respect to race, the proportion of fatalities among blacks in the state of Maryland is higher (29 percent) than the population ratio would predict (17.8 percent). This higher mortality does not come from Baltimore city where 60 percent of the state's black population lives and where black and white fatalities very nearly correspond to the population ratio. Rather, it is due to the contributions from the small counties where the black fatality rate is substantially greater than expected. This increase cannot be ascribed to cigarette smoking or alcohol intake, behavioral habits which are not notably different than in the other areas of Maryland. One can speculate that the higher fatality rate among rural blacks is due to economic reasons, reflected in the malfunctioning or unsafe use of mechanical equipment (such as stoves) with lower ignition safety or of furnishings and building materials with a tendency to rapid flame spread and increased intensity once ignition has taken place.

The high fatality incidence among black children (Figs. 3*a,b*) is noteworthy. This could, in part, be due to the absence of adult supervision, which prevents rescue from fires that are caused by mechanical malfunctions or by children playing with matches. By contrast, black fatalities in the 11 to 20 age group are unusually low.

TABLE 4—Actual versus expected fatalities based on age distribution
(expressed in percentages within a given age group).

| Age Group, years | Expected Incidents, based on census | Actual Incidents | | | | | | | |
|---|---|---|---|---|---|---|---|---|---|
| | | Maryland | Baltimore | 4 Large Counties | 18 Small Counties | Male | Female | White | Black |
| 0 to 2 | 2 | 4 | 6 | 4 | 2.5 | | | | |
| 0 to 10 | 18 | 22.5 | 27.7 | 23.7 | 15.9 | 17.2 | 28.9 | 15.2 | 40.9 |
| 11 to 20 | 20 | 11.7 | 9.2 | 10.5 | 15.9 | 13.1 | 10.0 | 15.5 | 2.3 |
| 21 to 30 | 15 | 11.7 | 11.6 | 15.8 | 5.4 | 7.6 | 16.5 | 12.0 | 10.5 |
| 31 to 40 | 11.8 | 6.6 | 14.1 | 2.3 | 5.4 | 8.8 | 4.0 | 6.8 | 6.6 |
| 41 to 50 | 10 | 12.4 | 11.6 | 10.5 | 15.9 | 17.2 | 6.5 | 12.3 | 12.4 |
| 51 to 60 | 10 | 17.5 | 14.1 | 21 | 15.9 | 20.0 | 14.7 | 19.0 | 17.5 |
| >60 | 13 | 17.5 | 11.6 | 15.8 | 26.4 | 16.2 | 19.4 | 19.1 | 17.5 |

## Alcohol Contribution

While specific contributions of alcohol to each individual case cannot be assessed because detailed information is unavailable, its overall influence on and correlation with the fire fatalities record is substantial. Ingestion of alcohol not only enhances the human errors that are the causes of most fires, but it makes escape more difficult. It depresses arousal to danger and slows down the taking of adequate countermeasures. It may also lower the lethality threshold of many toxic substances although there are no firm medical data to bear this out.

The contribution of alcohol is shown in Tables 5 and 6 and Figs. 5a,b, 6, and 7. The magnitude of involvement is expressed in terms of blood alcohol level. A value in excess of 0.1 percent corresponds to intoxication.

The seriousness of the alcohol problem is shown clearly in Fig. 5. Blood alcohol content above 0.1 percent rises sharply in victims beyond the age of 30 and only declines somewhat at ages greater than 60. This excessive alcohol intake is particularly prevalent for men.[6] Fifty-seven percent of all male fatalities involve alcohol, as against 26 percent for women (Fig. 6). In 71 percent of all fatalities in the age group 51 to 60, a connection with alcohol intake is observed. Alcohol is involved in 48 percent of all the white and in 39 percent of the black fatalities (Fig. 6). Forty-three percent of all the fire victims in Maryland had a positive test for alcohol. This rises to 51 percent if fatalities under 20 years of age are excluded.

By sex and race, the heaviest correlation of alcohol use and fire deaths comes from black men (where 62 percent of all fatalities involve alcohol), closely followed by white men (56 percent), and white women (35 percent). In contrast, no fire deaths involving alcohol have been recorded in black women (Fig. 6). Very heavy drinking (with blood alcohol levels in excess of 0.2 percent) is concentrated in Baltimore city (Fig. 7), with the small and large counties at a lower level. Statewise, it involves white males most (28 percent), followed by black males (26 percent). Heavy drinking among white women is low (3 percent) and entirely absent in black women.

## Time of Incidents

Fatal fire accidents occur throughout but are not uniformly distributed during the year or the time of day (Fig. 8 and Table 7). A preponder-

---

[6] The data from a report on alcohol drinking habits (Calahan, D., Cisin, I. M., and Crossley, H. M., "American Drinking Practices: A National Survey of Behavior and Attitudes Related to Alcoholic Beverages," Report 3, Social Research Group, The George Washington University, June 1967) do not show clear evidence that heavy drinking increases substantially with age for either men or women. It remains constant (at 25 ± 5 percent of the surveyed population) except for a decline beyond age 65. However, women are only about one quarter as much involved in heavy drinking as are men.

TABLE 5—*Blood alcohol level by age and location (1972/1973).*

| Location | Blood Alcohol, % | Age, years | | | | | | | |
|---|---|---|---|---|---|---|---|---|---|
| | | 0 to 10 | 11 to 20 | 21 to 30 | 31 to 40 | 41 to 50 | 51 to 60 | >60 | Sum |
| Maryland | negative | 23 | 7 | 6 | 3 | 4 | 6 | 8 | 57 |
| | ≤0.1 | 1 | 3 | 2 | 0 | 1 | 1 | 2 | 10 |
| | 0.1 to 0.2 | 0 | 1 | 2 | 3 | 1 | 4 | 4 | 15 |
| | >0.2 | 0 | 0 | 1 | 2 | 6 | 5 | 3 | 17 |
| | | 24 | 11 | 11 | 8 | 12 | 16 | 17 | 99 |
| Baltimore | negative | 12 | 3 | 2 | 1 | 1 | 5 | 3 | 27 |
| | ≤0.1 | 0 | 1 | 0 | 0 | 0 | 0 | 0 | 1 |
| | 0.1 to 0.2 | 0 | 0 | 1 | 3 | 0 | 0 | 1 | 5 |
| | >0.2 | 0 | 0 | 1 | 2 | 4 | 1 | 2 | 10 |
| | | 12 | 4 | 4 | 6 | 5 | 6 | 6 | 43 |
| 4 large counties | negative | 8 | 2 | 3 | 1 | 2 | 0 | 2 | 18 |
| | ≤0.1 | 1 | 1 | 2 | 0 | 1 | 0 | 2 | 7 |
| | 0.1 to 0.2 | 0 | 1 | 1 | 0 | 0 | 3 | 2 | 7 |
| | >0.2 | 0 | 0 | 0 | 0 | 1 | 4 | 0 | 5 |
| | | 9 | 4 | 6 | 1 | 4 | 7 | 6 | 37 |
| 18 small counties | negative | 3 | 2 | 1 | 1 | 1 | 1 | 3 | 12 |
| | ≤0.1 | 0 | 1 | 0 | 0 | 0 | 1 | 0 | 2 |
| | 0.1 to 0.2 | 0 | 0 | 0 | 0 | 1 | 1 | 1 | 3 |
| | >0.2 | 0 | 0 | 0 | 0 | 1 | 0 | 1 | 2 |
| | | 3 | 3 | 1 | 1 | 3 | 3 | 5 | 19 |

TABLE 6—*Blood alcohol level (99 cases) by sex, race, and location (1972/1973).*

| Location | Blood Alcohol, % | Male White | Female White | Male Black | Female Black | Male | Female | White | Black | Sum |
|---|---|---|---|---|---|---|---|---|---|---|
| Maryland | negative | 17 | 19 | 8 | 13 | 25 | 32 | 36 | 21 | 57 |
| | ≤0.1 | 4 | 4 | 2 | 0 | 6 | 4 | 8 | 2 | 10 |
| | 0.1 to 0.2 | 6 | 5 | 4 | 0 | 10 | 5 | 11 | 4 | 15 |
| | >0.2 | 11 | 1 | 5 | 0 | 16 | 1 | 12 | 5 | 17 |
| | | 38 | 29 | 19 | 13 | 57 | 42 | 67 | 32 | 99 |
| Baltimore | negative | 8 | 4 | 6 | 9 | 14 | 13 | 12 | 15 | 27 |
| | ≤0.1 | 1 | 0 | 0 | 0 | 1 | 0 | 1 | 0 | 1 |
| | 0.1 to 0.2 | 2 | 1 | 2 | 0 | 4 | 0 | 3 | 2 | 5 |
| | >0.2 | 6 | 0 | 4 | 0 | 10 | 0 | 6 | 4 | 10 |
| | | 17 | 5 | 12 | 9 | 29 | 14 | 22 | 21 | 43 |
| 4 large counties | negative | 6 | 9 | 0 | 3 | 6 | 12 | 15 | 3 | 18 |
| | ≤0.1 | 2 | 3 | 2 | 0 | 4 | 3 | 5 | 2 | 7 |
| | 0.1 to 0.2 | 4 | 3 | 0 | 0 | 4 | 3 | 7 | 0 | 7 |
| | >0.2 | 3 | 1 | 1 | 0 | 4 | 1 | 4 | 1 | 5 |
| | | 15 | 16 | 3 | 3 | 18 | 19 | 31 | 6 | 37 |
| 18 small counties | negative | 3 | 6 | 2 | 1 | 5 | 7 | 9 | 3 | 12 |
| | ≤0.1 | 1 | 1 | 0 | 0 | 1 | 1 | 2 | 0 | 2 |
| | 0.1 to 0.2 | 0 | 1 | 2 | 0 | 2 | 1 | 1 | 2 | 3 |
| | >0.2 | 2 | 0 | 0 | 0 | 2 | 0 | 2 | 0 | 2 |
| | | 6 | 8 | 4 | 1 | 10 | 9 | 14 | 5 | 19 |

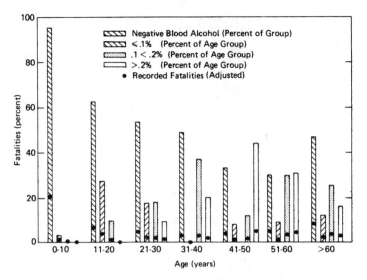

FIG. 5a—*Correlations of fire fatalities in Maryland as function of age.*

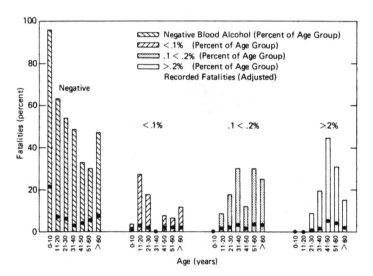

FIG. 5b—*Correlations of fire fatalities in Maryland as function of blood alcohol content.*

ance of fatal accidents takes place between midnight and 6 a.m. (42.5 percent), followed by 6 p.m. to midnight (22.1 percent), 6 a.m. to noon (18.6 percent), and noon to 6 p.m. (16.8 percent). The midnight to 6 a.m. period involves adults slightly more (44 percent) than children (36 percent). Also, alcohol-related fatalities peak in that period, but the fraction

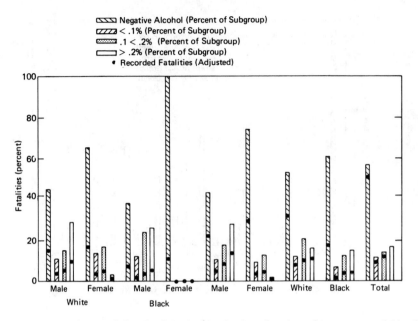

FIG. 6—*Correlation of fire fatalities in Maryland as function of sex, race, and blood alcohol content.*

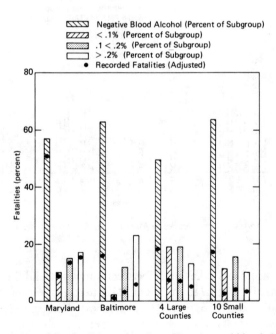

FIG. 7—*Correlation of fire fatalities as function of location and blood alcohol content.*

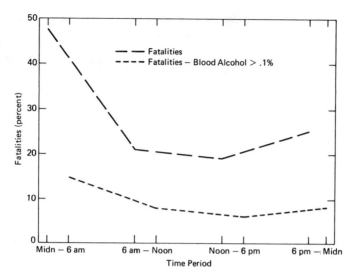

FIG. 8—*Correlation of fire fatalities as function of time of day.*

TABLE 7—*Occurrence of fatal fire accidents by time of day
(113 cases, not adjusted, 1972/1973).*

| Time | Total | Adults | Adults, alcohol >0% | Children (<10 years old) | Children (<2 years old) |
|------|-------|--------|---------------------|--------------------------|-------------------------|
| Midnight to 6 a.m. | 48 (41%) | 39 (44%) | 15 | 9 (36%) | 2 |
| 6 a.m. to noon | 21 (18.6%) | 15 (17%) | 8 | 6 (24%) | 3 |
| Noon to 6 p.m. | 19 (16.8%) | 14 (15.9%) | 6 | 5 (20%) | 3 |
| 6 p.m. to midnight | 25 (22.1%) | 20 (22.7%) | 8 | 5 (20%) | 2 |
| | 113 | 88 | 37 | 25 | 10 |

of intoxicated fire victims is highest during the day. Children under the age of two are most heavily involved in the period 6 a.m. to 6 p.m., indicating that the absence of adults to assist in rescue may be of importance.

With regard to seasonal distribution, the six cold months of the year (October to March) are responsible for 69 percent of all fatalities (Table 8 and Fig. 9). Alcohol involvement is also heaviest during this period (69 percent), as are the number of incidents in which more than one person is killed (71.7 percent).

## Physical Causes of Fires

Reconstruction of the events that initiated fires with fatal outcomes depends on detailed investigation at the fire scene. The causes can be

TABLE 8—*Occurrence of fatal fires by time of year (1972/1973 not adjusted).*

| Month | Total Fatalities | Multiple Fatalities | Alcohol >0.1% in Total Fatalities |
|-------|------------------|---------------------|------------------------------------|
| January | 20 | 12 | 3 |
| February | 8 | 2 | 5 |
| March | 8 | 4 | 4 |
| April | 5 | 0 | 2 |
| May | 11 | 6 | 5 |
| June | 2 | 0 | 1 |
| July | 0 | 0 | 0 |
| August | 11 | 5 | 3 |
| September | 2 | 0 | 0 |
| October | 10 | 0 | 4 |
| November | 8 | 4 | 4 |
| December | 15 | 6 | 5 |
| | 100 | 39 | 36 |

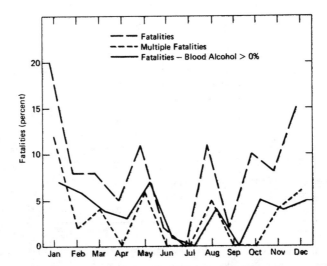

FIG. 9—*Correlation of fire fatalities as function of time of year.*

divided into two broad categories: (*a*) human errors or deliberate fire starts and (*b*) malfunction of mechanical equipment.

Table 9 presents a listing of the most probable causes. The importance of cigarettes is evident. A sizeable percentage (47.2 percent) of the fatal fires had their origin in the careless handling of cigarettes. The remainder were spread among a wide variety of human misjudgments (careless use of matches and of flammable liquids) or were due to faulty equipment (electrical short-circuits, etc.). Motor vehicle and airplane crashes or

TABLE 9—*Physical causes of fatal fires as indicated by fire investigators.*

| | |
|---|---|
| Smoking | 50 |
| Matches | 5 |
| Candles | 2 |
| Cooking | 3 |
| Flammable liquids | 2 |
| Heaters | 13 |
| Electrical malfunctions | 3 |
| Suicides | 3 |
| Suspicious origins | 9 |
| Unknown | 9 |
| Total | 99 |

explosions, where ensuing fires were judged to be the primary cause of death, were rare or nonexistent.

Since cigarettes are the major cause of fire fatalities, it is of interest to analyze the available data in more detail (Table 10). Sixty-one percent of the Baltimore city fatalities involve cigarettes, as compared to 50 percent and 32.6 percent in the large and small counties, respectively, indicating a substantially larger use and misuse of cigarettes in the city and the suburbs. Almost half (42 percent) of the cigarette-caused ignitions occur in the midnight to 6 a.m. period. The other periods are almost equally involved at half this rate.

As one would suspect, excessive alcohol ingestion and cigarette-caused

TABLE 10—*Cigarettes as ignition source in fire fatalities.*

| | Male White | Female White | Male Black | Female Black | Male | Female | White | Black | Sum |
|---|---|---|---|---|---|---|---|---|---|
| Total fatalities (unadjusted)[a] | 24 | 14 | 7 | 5 | 31 | 19 | 38 | 12 | 50 |
| Blood alcohol > 0% (unadjusted) | 16 | 8 | 7 | 0 | 23 | 8 | 24 | 7 | 31 |

| Location | No Alcohol | Alcohol > 0% | Cigarette Caused Fatalities | Total Fatalities | Cigarette Caused Fatalities, % |
|---|---|---|---|---|---|
| Maryland (adjusted)[b] | 14.2 | 27.7 | 42.5 | 88.7 | 47.9 |
| Baltimore (adjusted) | 7.0 | 8.1 | 15.1 | 24.9 | 60.6 |
| 4 large counties (adjusted) | 5.0 | 14.0 | 19.0 | 38.0 | 50.0 |
| 18 small counties (adjusted) | 2.8 | 5.6 | 8.4 | 25.8 | 32.6 |

[a](unadjusted)—total case load.
[b](adjusted)—corrected for differences in participation in the fire fatalities program among the three major jurisdictions (Baltimore city, 4 large counties, 18 small counties).

ignition are closely linked. In two thirds of the fatalities ascribed to cigarettes (that is, 31 in 50) elevated alcohol contents were found. Further details are given in Table 10. The larger contribution of men to cigarette-caused deaths and their heavy involvement in the smoking/drinking link is evident.

## Medical Causes of Fatal Fires

The fatal outcome of fires on human beings can be due to either a single or to a combination of distinct classes of causes. One class is made up of a broad spectrum of chemicals that are produced and inhaled at the fire scene. During combustion, many compounds are generated which are highly toxic to human beings.[7] It is difficult to predict their nature and concentration in detail since these factors are strongly dependent on the manner by which fires engulf the burnable components, on their composition, and on the gas flow pattern within the burning structure.[8] Some of the substances are potential asphyxiants (such as carbon monoxide (CO)[9] which displaces oxygen from hemoglobin and thus disturbs the normal oxygen delivery in the body). Others are specific in their interference with specific enzymatic reactions (hydrogen cyanide (HCN)). Acidic or irritating gases (such as hydrochloric acid (HCl) from the combustion of polyvinyl chloride[10] or aldehydes) may lead to severe pulmonary disorders. Combustion products containing highly toxic agents affecting the central nervous system[11] and metal-containing soots have recently been identified.[12]

---

[7] Kishitani, K. and Nakamura, K., "Toxicity of Combustion Products," *Journal of Fire and Flammability/Combustion Toxicology,* Vol. 1, 1974, p. 104. Kishitani, K., "Study of Injurious Properties of Combustion Products of Building Materials at Initial Stages of Fire," *Journal of the Faculty of Engineering,* University of Tokyo (B), Vol. 31, No. 1, 1971. Einhorn, I. N., "Physiological and Toxicological Aspects of Smoke Produced During the Combustion of Polymeric Materials," *Environmental Health Perspectives,* Vol. 11, 1975, p. 163.

[8] Madorsky, S. L., *Thermal Degradation of Polymers,* Reinhold, New York, 1968. Wagner, J. P., "Survey of Toxic Species Evolved in the Pyrolysis and Combustion of Polymers," *Fire Research Abstracts and Reviews,* Vol. 14, 1972, p. 1. Wooley, W. D., *Plastics and Polymers,* Dec. 1973, p. 280. Bowes, P. C., "Smoke and Toxicity Hazards of Plastics in Fires," *Annals of Occupational Hygiene,* Vol. 17, 1974, p. 143.

[9] Haldane, J. S. and Priestley, J. G., *Respiration,* Yale, 1935. Drinker, C. K., *Carbon Monoxide Asphyxia,* Oxford, England, 1938. Killick, E. M., *Physiological Review,* Vol. 20, 1940, p. 313. Meigs, J. W. and Pugh, J. G. W., *Archives of Industrial Hygiene,* Vol. 6, 1952, p. 344. *Carbon Monoxide: A Bibliography With Abstracts,* U.S. Department of Health, Education and Welfare, Public Health Service, 1966. Bour, H. and Cedingham, I. McA., Eds., "Carbon Monoxide Poisoning," *Progress in Brain Research,* Vol. 24, Elsevier, New York, 1957. "Biological Effects of Carbon Monoxide," *Annals of the New York Academy of Sciences,* Vol. 174, 1970, p. 1.

[10] Coleman, E. H. and Thomas, C. H., "Products of Combustion of Chlorinated Plastics," *Journal of Applied Chemistry,* Vol. 4, 1954, p. 379. Tsuchiva, Y. and Sumi, K.,

A second class covers the effects of unusual stress on persons involved in fires. Heart attacks or severe hormonal upsets can have fatal outcomes, especially if aggravated by disturbances brought on by the exposure to noxious chemicals.

A third broad class includes thermal causes, of which burns are the most common example. Accidents, such as falls, cuts, or bruises incurred during escape from fire situations, may also lead to fatal outcomes.

In this study, detailed biochemical analyses were made as part of a general autopsy to identify the principal causes of the deaths.

Initially, the primary measurements were confined to carboxyhemoglobin (COHb) levels as indication of CO ingestion. At a later stage, lungs of fire victims were analyzed for volatile organic gases[13] and for soot deposits. Most recently, blood cyanide levels have been measured as well.[14] In addition, an investigation of CO interaction with preexisting heart malfunction is under way.[15] Observations on the extent of burns are also available.

---

"Thermal Decomposition Products of Polyvinyl Chloride," *Journal of Applied Chemistry,* Vol. 17, 1967, p. 364. Stone, J. P., Hazlett, R. N., Johnson, J. E., and Carhart, H. W., "The Transport of Hydrochloric Acid from Burning Polyvinyl Chloride," *Journal of Fire and Flammability,* Vol. 4, 1973, p. 42. The presence of hydrochloric acid condensed on or absorbed by soots is frequently observed during the analysis of solid combustion products found at the scenes of fires (Berl, W. G., *Proceedings,* First Conference and Workshop on Fire Casualties, Applied Physics Laboratory, The Johns Hopkins University, April 1976, p. 73). The detection in soots deposited in the tracheobronchial trees of fire casualties is masked by the chloride ion content of biological fluids. The possibility of severe, local chemical irritations of lung tissue is obvious.

[11] Petajan, J. H., Voorhees, K. J., Packham, S. C., Baldwin, R. C., Einhorn, I. N., Grunnet, M. L., Dinger, B. G., and Birkey, M. M., "Extreme Toxicity from Combustion Products of a Fire-Retarded Polyurethant Foam," *Science,* Vol. 187, 1975, p. 742.

[12] Soots ingested by fire casualties show a surprisingly frequent presence of metals, often in relatively high concentrations. Lead (presumably from paint pigments), copper, and cadmium (from electrical wiring and fixtures), and antimony (from fire inhibitors) have been detected in nearly one quarter of all fire victims. While their contribution to the fatal outcomes appear remote, the significance of these findings for people recovering from exposure to other combustion products remains to be established.

[13] Evidence for irritating organic compounds has been obtained by analysis of lung gases. Acetaldehyde is present in detectable amounts in the lungs of most fire fatalities (to be published).

[14] The development of a reliable analytical method for hydrogen cyanide in human blood (Altman, R., *Proceedings,* First Conference and Workshop on Fire Casualties, Applied Physics Laboratory, The Johns Hopkins University, April 1976, p. 65) has led to routine blood screening of fire fatalities. The significance of its frequent presence and its contribution to debilitation and death in the presence of carbon monoxide needs clarification.

[15] A surprisingly large number of fire victims have been shown to have had preexisting heart disease (Fisher, R. S., *Proceedings,* First Conference and Workshops on Fire Casualties, Applied Physics Laboratory, The Johns Hopkins University, April 1976, p. 41), particularly in the age span 20 to 39. While a relationship of oxygen deprivation and heart failure has been postulated before (see Comroe, J. H., *Physiology of Respiration,* Year Book Medical Publishers, 1974), the availability of a large number of fire victims with varying degrees of carbon monoxide ingestion will make it possible to establish this suspected relationship more firmly.

Figure 10 presents a summary of the biochemical measurements of the COHb content in the blood of fire fatalities, as well as the blood alcohol content and an indication of ignition caused by cigarettes. The results are shown in Table 11. If a COHb content of 50 percent or higher is set as the lethal limit, the majority of the casualties (59 percent) had inhaled lethal quantities of CO. Other debilitating compounds may have been involved also, for which chemical analyses were not made. In another group of eleven fatalities, the CO content was sufficiently high (that is, above 30 percent COHb) to have caused severe debilitating reactions.

## Discussion

Despite the unquestioned complexity of the physical and medical causes that lead to death in fires and the difficulty of obtaining adequately detailed

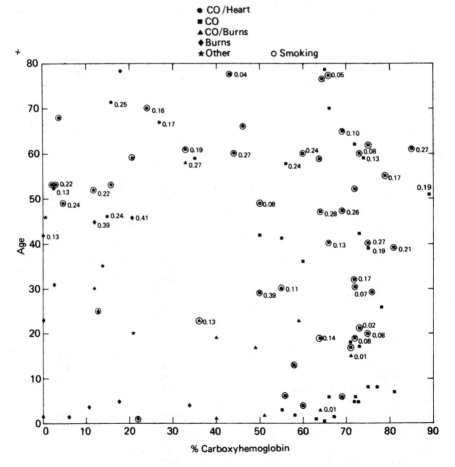

FIG. 10—*Correlation of age, blood carboxyhemoglobin content, most probable medical causes of death, and smoking as initiator of fatal fires. Numbers refer to blood alcohol levels.*

TABLE 11—*Blood carboxyhemoglobin levels of fire fatalities.*

| COHb, % | Total | Children Under 10 years | 10 to 60 Years Age Group | Adults Above 60 years | Alcohol >0% | Heart Condition | Other Causes of Fatality |
|---|---|---|---|---|---|---|---|
| 0 to 10 | 11 | 2 | 8 | 1 | 4 | 3 | 1 spleen injury |
| 11 to 20 | 11 | 2 | 7 | 2 | 4 | 3 | 1 bleeding |
| 21 to 30 | 6 | 1 | 3 | 2 | 3 | 2 | |
| 31 to 40 | 7 | 2 | 5 | 0 | 3 | 1 | |
| 41 to 50 | 7 | 0 | 5 | 2 | 4 | 2 | |
| 51 to 60 | 12 | 5 | 7 | 0 | 3 | not available | |
| 61 to 70 | 16 | 6 | 5 | 5 | 7 | not available | |
| 71 to 80 | 25 | 5 | 18 | 2 | 11 | not available | |
| >80 | 4 | 1 | 2 | 1 | 3 | not available | |
| | 99 | 24 | 60 | 15 | 42 | | |

information and reconstructing the important chains of events, it is possible to reach a number of general conclusions that shed some light on the major problem areas. Interpretations may change with time since they are based on *ex post facto* evaluations of potentially varying human behavioral and technological factors. But, insofar as Maryland is typical of the living conditions and life styles of many parts of the United States, the results may be suggestive of the general problem elsewhere.

Most fire fatalities are solitary incidents or incidents that involve only a few persons. They occur primarily in residential structures and are largely the result of human errors. Few fire deaths are deliberately planned or self-inflicted. Spectacular and costly disasters in schools, movies, airplanes, and public meeting places are rare and hardly affect the national annual fatality records. During the entire four-year span of the fatality study, no mass disasters were encountered.

Also, the generally quoted substantial contribution of automobile crash fires to the death toll[16] was almost entirely absent when the medical results of the investigations were analyzed in detail. This is so contrary to the accepted statistical information of nearly one third of all fire fatalities occurring in automobile fires that a detailed check on the reasons for this discrepancy is in order.

All age groups are involved in fire fatalities, with children under ten years of age and adults beyond the age of forty involved well beyond their proportional number in the population. Black children in Baltimore city and old people in the small counties are particularly vulnerable groups. In both cases, insufficient help with escape during the emergency may be an explanation.

Men make up a substantially larger fraction of the fire casualties than women. This is related to a marked increase in the ingestion of alcohol by men in the age groups beyond 30. Most likely, it has an effect on the number of accidental ignitions and on the ability to take proper countermeasures.

Fire fatalities have occurred in nearly all of the political subdivisions of the state of Maryland. The average per capita record in the smaller, rural counties and in urban Baltimore is almost twice as poor than that in the larger, suburban counties. It is difficult, on the basis of the available information, to provide an explanation for this difference. Behavior patterns, standards of building construction, and fire safety of furnishings vary considerably with the various locations. One can speculate that the response time of local fire departments may have an effect in reducing the casualties in the large counties. Medical treatment may also be superior in the suburban communities. However, to confirm such tentative explana-

---

[16] The Report of the National Commission on Fire Prevention and Control (*America Burning*) presents fire loss data (Appendix V) for 1971. The life loss in building fires is given as 7570. Motor vehicle-caused fire losses are 3950.

tions would require a much more refined analysis of the fire development, of human behavior, and the effectiveness of countermeasures. Such detailed information does not currently exist.

It is tempting to speculate about the wide variations in COHb content of fire victims and their relationship to the manner in which death occurred. The majority of the 59 casualties showing COHb values in excess of 50 percent was probably due to slow ingestion of toxic gases (mainly CO) which were generated by smoldering combustion.[17] Small quantities of a carbonaceous fuel (in the order of one tenth of a pound) can generate enough carbon monoxide to produce a concentration in excess of 0.1 percent CO in a room of 1600 ft$^3$ volume (10 by 20 by 8 ft), with negligibly small temperature rise or oxygen depletion. One would expect high levels of carbon monoxide to be the sole cause of death.

Should smoldering combustion be followed by a free-burning but ventilation-controlled fire, it is conceivable that moderate COHb concentration (30 to 50 percent), together with burns, are the joint causes responsible for the fatal outcomes. In such cases, the initial ingestion of CO occurred in the absence of a high-temperature-producing flame. By itself, this CO ingestion would have been sufficient to cause death but could have led to severe incapacitations. But at a later time, the free-burning fire has advanced toward the victim close enough so that serious burns were caused. Evasive action was made difficult or impossible by the earlier COHb buildup.

In the case of fire victims with COHb contents below 20 percent, fatalities in most instances were due to the effects of rapidly evolved hot gases from explosions or flash fires. This can lead to very high temperatures and extensive oxygen depletion in a short time. Carboxyhemoglobin buildup is either negligible or low, partly because the CO content in such fuel-controlled fires is low, partly because the high gas temperature leads to rapid stoppage of breathing (laryngeal spasms). Fuel vapor explosions, rapidly progressing fires in compartments, or rapidly propagating clothing fires are representative examples.

An unexpectedly large number of fire victims in the low COHb range were found to have suffered with severe cardiac difficulties, amounting to 9 out of the 34 fatalities with COHb contents of less than 40 percent. Since these fatalities were also strongly linked with cigarette smoking as a cause of the fire start (Table 10), it is conceivable that, in some instances, a fatal heart attack may have been the cause rather than the consequence of the fire. It is not possible, at this time, to judge the importance of other eventualities such as the effects of stressful situations

[17] Conditions leading to smoldering or flaming combustion are described in Palmer, K. N., Taylor, W., and Paul, K. T., "Fire Hazards of Plastics in Furniture and Furnishings: Characteristics of the Burning," Building Research Establishment Current Paper CP 3175, Fire Research Station, Borehamwood, England, Jan. 1975.

during a fire or the combination of a moderate CO ingestion and heart insufficiency, particularly when coupled with high alcohol intake.

## Conclusions

Fire deaths are distributed throughout the state of Maryland, but the rate is notably lower (approximately one half) in suburban communities as compared with Baltimore city or rural areas. Deaths in children and the elderly occur well beyond their proportion in the population. Fire disasters, involving many people, and fatal transportation fires were rare in this two-year study and did not contribute much to the total fatality record.

Inhalation of CO is responsible for incapacitation and death of 72 percent of the fire fatalities. A pronounced involvement of alcohol (detected in 43 percent of all fire victims), particularly with men between the ages of 40 and 60, has been observed, coupled with careless use of cigarettes.

Warning devices sensitive to smoke or other products of combustion could have prevented many of the fatalities in which high carbon monoxide intakes were involved. An effective public information effort, pointing up the hazards to life of simultaneous smoking and excessive drinking, should prove beneficial. Large life hazards posed by cigarettes as ignition source for fatal fires have been observed. Improved materials and designs that prevent smoldering, cigarette modifications, and public education on the hazards represent appropriate countermeasures.

*Acknowledgments*

The following organizations made major contributions in the acquisition of the data on which this study is based: Office of the Medical examiner, State of Maryland; Maryland State Fire Marshal; Baltimore City Fire Department; Johns Hopkins University School of Hygiene and Public Health; U.S. National Bureau of Standards; Fire Problems Program, Applied Physics Laboratory, The Johns Hopkins University.

# DISCUSSION

---

*J. H. Petajan*[1] (*oral discussion*)—In the statistics on fatalities caused by cigarettes, how many of the victims were "primary," that is responsible for the ignition, and how many were "secondary," that is, not responsible for the initiation of the fatal fire?

[1] University of Utah, Salt Lake City, Utah 84112.

*W. G. Berl* (*author's oral closure*)—Of the 50 recorded fatalities due to cigarette-ignited fires (in a total sample of 100 casualties), 30 were due to ignitions in which the cigarette smoker became the single fatality. In 2 cases, the solitary victim did *not* cause the ignition. In 9 multiple incidents, 18 persons died, of whom 9 were responsible for the cigarette-caused ignition. Thus, 11 fatalities were "secondary" in the sense that the victims bore no responsibility for the fire cause. Of these, 4 were less than 10 years old.

One last point I would like to make. We find it very difficult, on the basis of what is happening in the state of Maryland, to account for an annual fatality of 12 000 people. We rather think it is more like six or seven thousand, extrapolating Maryland to the country as a whole. This is a serious matter yet to be resolved.

*C. F. Klein*[2] (*written discussion*)—What are the long-term effects of exposures to low levels of toxic gases, such as CO, HCN and HCl?

*W. G. Berl* (*author's written closure*)—The effects of repeated exposures to toxic gases at sublethal levels on humans are difficult to determine. An effort is currently underway in "Project Smoke," carried out by the Johns Hopkins University School of Hygiene and Public Health to analyze the long-term effects of exposures to fire atmospheres and stresses on fire fighters. This study, however, cannot give insights into the contributions from individual gases.

*D. Hammerman*[3] (*written discussion*)—Has an effort been made to collect data on blood COHb levels in survivors who were nearby or in the fire area where a fire fatality occurred?

*W. G. Berl* (*author's written closure*)—No systematically acquired information on COHb is available on people that have been exposed to fire atmospheres. Generally, if toxic effects from high intakes of carbon monoxide are suspected, oxygen is administered as quickly as possible in order to reverse the debilitating effects of CO.

It would be of value, however, to gather more information on survivors from fires who have ingested toxic combustion products. It would help to distinguish effects that may not be caused by CO (such as hyperventilation) and for which oxygen therapy is not a cure. There is also little factual information on hand on sublethal effects of CO that cause severe physiological and psychological malfunctions in stressful situations.

*C. A. Clark*[4] (*written discussion*)—Can the data be analyzed in terms of deaths: (*a*) where the victim was known to have been in the room of fire origin and (*b*) where the victim was not in the room of fire origin?

*B. M. Halpin* (*author's written closure*)—In the analysis of the data,

---

[2] Senior research engineer, Johnson Controls, Inc., Milwaukee, Wis. 53201.

[3] Chief fire protection engineer, Office of the Maryland Fire Marshal, Baltimore, Md. 21201.

[4] Senior development scientist, B. F. Goodrich Chemical Co., Avon Lake, Ohio 44012.

the room of fire origin and the location where the victim is found are both recorded. Thus, one can determine whether or not the victim was found in the room of origin.

In about 40 percent of the cases, the victim was found in the room where the fire originated. However, it is not always discernible whether the victim was there at the time that the fire started. Some persons clearly entered the room of fire origin on a search mission. The details of such special situations are very difficult to reconstruct.

Similarly, many victims escaped from the room of fire origin only to die somewhere else. Again, it is not possible to retrace their movements in detail.

*J. R. Bercaw,*[1] *K. G. Jordan,*[1] *and A. Z. Moss*[1]

# Estimating Injury from Burning Garments and Development of Concepts for Flammability Tests

**REFERENCE:** Bercaw, J. R., Jordan, K. G., and Moss, A. Z., "**Estimating Injury from Burning Garments and Development of Concepts for Flammability Tests,**" *Fire Standards and Safety, ASTM STP 614*, A. F. Robertson, Ed., American Society for Testing and Materials, 1977, pp. 55–90.

**ABSTRACT:** Part of the flammability research at Du Pont has been concerned with understanding the way various garments ignite and transfer heat under simulated conditions of use. The objective is to develop basic principles which will aid in the definition of improved products and guide the development of more appropriate test methods. To this end, we have acquired Thermo-Man, a thermally instrumented manikin and heat sensing system which produces detailed estimates of injury from burning garments. The highly sophisticated system and procedure for injury evaluation to give accurate estimates of area burned and burn depth through the human skin is reviewed. Burn injury data are presented which predict injury potential from common apparel, including fabrics from both man-made and natural fibers. Effects of garment type on potential injury and injury-time propagation are also discussed.

Thermo-Man provides a unique basis for development of concepts for more relevant and predictive flammability tests applicable to general apparel fabrics. The essential aspects of burn injury production are fabric ignition and heat transfer. Research at the National Bureau of Standards (NBS) and Du Pont is aimed at developing a heat transfer concept for possible consideration as part of a general apparel fabric flammability test method.

**KEY WORDS:** fires, flammability, flammability testing, fabrics, apparel fabrics, heat transfer, burns (injuries)

The basic objective of product safety tests and standards is to protect the public from the unreasonable risk of serious injury, death, or property loss. In addition, to protect the public from undue burden, such standards should be addressed to those areas where:

1. Need is the greatest.

[1]Manager, Product Services and Technology, Textile Fibers Dept., senior research chemist, Textile Research Laboratory, and research chemist, Textile Research Laboratory, respectively, E. I. du Pont de Nemours and Co., Inc., Wilmington, Del. 19898.

2. Hazard can be defined.
3. Relevant test method can be designed to protect against such hazard.
4. Reasonable cost-benefit can be predicted.

It is probably obvious by now that we have just reviewed many points stated in the Flammable Fabrics Act and voiced by the Consumer Product Safety Commission.

The purpose of this paper is to discuss some of our efforts at Du Pont as they relate to the second and third points just listed. Definition of any real-life flammability hazard is a highly complicated task. Considerable progress is being made using sophisticated facilities and instrumentation to better define and understand the combustion properties of materials for construction and home furnishing. The difficulties of such studies are, in many ways, even more complicated when we attempt to define any real-life hazards of apparel fabrics because of human involvement and interaction.

Textile flammability is a matter of real concern to Du Pont, as well as the industry in general. We want to develop a factual understanding of the problem so that we can take intelligent courses of action. As a result, part of our fabric flammability research at Du Pont has been concerned with understanding the way various garments ignite and transfer heat under realistic conditions. Our objective is to develop basic principles which will aid in the development of more appropriate test methods and in the definition of improved textile products.

**Equipment And Procedures**

Our Textile Research Laboratory has placed heavy emphasis on the measurement of heat transferred from burning fabrics and garments to a sensing surface, since heat transfer is a key factor in producing actual burn injury. Early work in this area was based on use of a fairly simple heat transfer panel and a simple instrumented child-sized manikin [1].[2] Much useful information was obtained from experiments conducted with these devices, but they also had some shortcomings. Recognizing the need for more quantitative research in this area, we acquired Thermo-Man, a highly sophisticated manikin and sensing system which is capable of producing detailed estimates of injury from burning garments. Thermo-Man was originally developed by the Aerotherm Division of the Acurex Corporation for U. S. Air Force studies of protective clothing.

The Du Pont Thermo-Man system shown in Fig. 1 was modified specifically to study general apparel flammability. The general principle of operation is for the heat sensors to detect temperature at a given time, which is then translated by a computer program into extent of body dam-

[2] The italic numbers in brackets refer to the list of references appended to this paper.

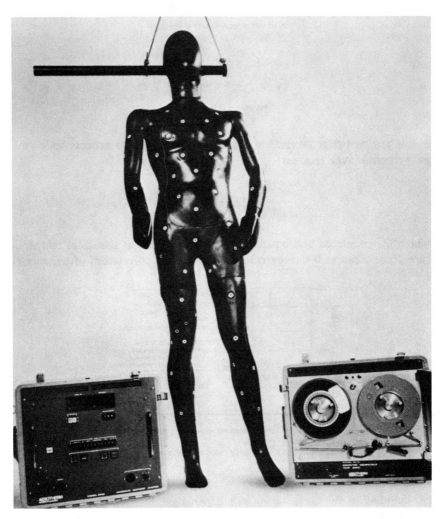

FIG. 1—*Thermo-Man.*

age. The procedure for injury evaluation is shown in Fig. 2. Each of Thermo-Man's 122 sensors (including 8 sensors in the face, neck, and head) can be scanned up to 3 times/s to record temperature-time history during the experiment. This is converted into incident surface heat flux which serves as the input to a skin-simulant computer program. Skin surface temperature history is computed, and heat transfer through the skin is determined using thermal properties of human skin. Knowing the temperature history as a function of depth allows depth of thermal injury to be determined. The burn injury model developed by Aerotherm follows the work of Henriques [2] which was modified by Stoll et al [3,4]. The

FIG. 2—*Procedure for injury evaluation.*

model assumes that thermal injury is an activated rate process following an Arrhenius-type relation

$$\frac{d\Omega}{d\theta} = \left( C \exp - \frac{\Delta E}{RT} \right)$$

where $\Omega$, integrated and equated to unity, means that tissues at temperature $T$ are dead in $\theta$ s. Figure 3 shows excellent correlation of the model

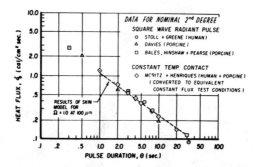

FIG. 3—*Correlation of skin burn tests with mathematical model of skin injury.*

(as depicted by the dashed line) with experimental second degree or partial thickness burn injuries on human and porcine subjects [2–7]. In our Thermo-Man system, this relationship has been extrapolated to estimate full-thickness or third degree burns, although its correlation with living skin cannot be so precisely established.

Modifications are being made periodically to our Thermo-Man to make it as predictive as possible of deep injuries. For example, we have changed the computer program to account for the liquid-to-vapor phase change that occurs at 100 °C for tissue water. Without this modification, our computer program overstates third-degree tissue damage. Also, a controlled wind system has been added to simulate Thermo-Man in motion (Fig. 4).

We believe that the Thermo-Man system provides an ability to study apparel flammability in a quantitative way heretofore not possible. It is

FIG. 4—*Thermo-Man chamber.*

the only one being used in the textile industry to study apparel flammability.

The actual burn damage information provided by Thermo-Man is described in Figs. 5, 6, and 7. Computer printouts include a summary of burn damage as percent body burned by class or degree of injury at various depths into the skin. This is measured in terms of $\Omega$, the Henriques' tissue damage factor. Class A means no epidermal damage, while Class B signi-

*I.*  SUMMARY OF BURN DAMAGE TO THE TOTAL BODY

- CLASS A - NO DAMAGE    $(\Omega < 0.5$ AT $100\mu)$
- CLASS B - 1st DEGREE    $(0.5 < \Omega < 1.0$ AT $100\mu)$
- CLASS C - 2nd DEGREE    $(\Omega = 1.0$ BETWEEN $100$ & $2000\mu)$
  PARTIAL THICKNESS
- CLASS D - 3rd DEGREE    $(\Omega > 1.0$ AT $2000\mu)$
  FULL THICKNESS

( WHEN $\Omega = 1$ , ALL CELLS ARE DESTROYED )

*II.*  BODY REGIONS

- % DAMAGE BY CLASS
- $\Omega$ AT $100\mu$ AND $2000\mu$
- HEAT ( CAL./CM² )

FIG. 5—*Thermo-Man burn damage information.*

*TIME - INJURY PROFILE*

- *NUMBER OF SENSORS REACHING  Ω = 1  AT 10*
  *DEPTH LEVELS  VS.  TIME*
- *TIME  VS.  CLASS  OF  INJURY*

*HEAT  TRANSFER  DATA*

- *TIME  VS.  TEMP.*
- *TEMP.  AND  Ω  AT  100μ*
- *HEAT  FLUX  &  TOTAL  HEAT  VS.  TIME*

FIG. 6—*Thermo-Man burn damage information (continued).*

fies first-degree burns or reddening of the skin. Class C indicates second-degree or partial-thickness burns through the epidermis which averages about 100 μm in depth. This is accompanied by blistering. Class D indicates third-degree or full-thickness burns through the dermis and corresponds to an injury depth which can exceed 2000 μm in depth. An artist's conception of human skin is shown in Fig. 7. Second- and third-degree (C and D) injury exceeding 15 to 20 percent of the body area normally requires prolonged hospitalization in a specially equipped burn-trauma unit. Since skin regenerating functions such as sweat and sebaceous glands and hair follicles are destroyed in third-degree burns, skin grafting is required.

The computer printout also breaks down body injury by body region burned. Complete heat transfer information is also provided. As shown in Fig. 6, a time versus degree-of-injury profile and heat transfer data, including skin surface temperature at any given time, tissue temperature and damage at 100 μm, and heat-flux plus total heat versus time, are printed out. We can also obtain a plot of body area burned as a function of time for each burn class.

Before illustrating the kinds of information which the Thermo-Man system can generate, we would like to caution against drawing too many general conclusions from the data. As you know, many factors affect apparel flammability and any related injuries. Among these factors are the garment design, fit, and condition, the fabric composition, construction, finishes, the conditions of ignition, the presence of flammable liquids, the victim's reactions, and many others. We can only present data for very simple accident circumstances. Every real incident is unique, and we do not presume to say that Thermo-Man duplicates real life. However, we do believe that Thermo-Man is the closest we can get today to simulating a real apparel flammability accident in a laboratory environment.

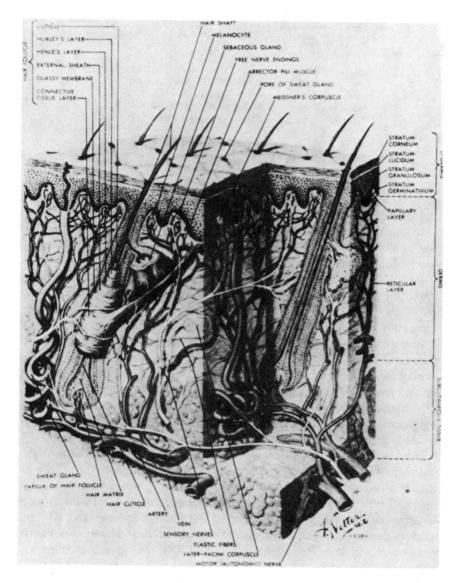

FIG. 7—*Human skin.*

## Experimental Work

Our experimentation with Thermo-Man has been underway for over a year. We have tested an extensive number of commercially available and experimental fabrics, some of which are shown in Fig. 8. As you can see, the list includes a wide range of 100 percent textured polyester wovens

- *100% TEXTURED POLYESTER*
  - *KNITS AND WOVENS*
    - *WEIGHT RANGE = 1 5 TO 7 6 OZ/YD$^2$*
    - *45 DIFFERENT FABRICS PURCHASED AT RETAIL*

- *COTTON, 3.2 OZ.*
- *POLYESTER / COTTON , 2.9 OZ*
- *ACRYLIC , 3.5 OZ*
- *WOOL , 3.5 OZ .*
- *ACETATE , 2.8 OZ .*

- *NYLON , 4.0 OZ .*
- *SPUN POLYESTER, 6.3 OZ .*
- *SPUN FR POLYESTER, 6.3 OZ .*
- *SPUN FR POLYESTER /FR RAYON, 4.3 OZ .*
- *FR COTTON , 4.9 OZ .*

FIG. 8—*Fabrics tested on Thermo-Man.*

and knits, untreated cotton, untreated polyester/cotton, acrylic, wool, acetate, nylon, spun polyester, spun flame-retarded polyester, blends of spun flame resistant (FR) polyester with FR rayon, and FR cotton.

In our first series of experiments, our intent was to define Thermo-Man's ability to distinguish among these various textile compositions, while keeping garment design and ignition source constant.

We, of course, recognize that ease of ignition and extinction are very important considerations in garment flammability properties. However, in our work to date, we have elected to emphasize the question: "once ignited, how much damage was caused by the burning garment?"

Single layer A-line garments with no undergarments were ignited with a triangular tab of filter paper weighing 1.1 g. It is attached at the hemline behind the right knee, as shown in Fig. 9. This paper tab size provides an exposure similar to that of a book of matches or a few-second exposure to a gas range burner. It is substantially more severe than the 3-s exposure in Federal flammability standards DOC FF 3-71 or FF 5-74 for children's sleepwear.

Burn injury data were computed during 80 s following ignition of the garment. We recognize that the first 30 to 60 s are probably the most critical in an actual apparel flammability accident. However, the 80-s duration and other conditions were selected after considerable experimentation to approximate injury results reported for similar apparel and ignition sources in the National Bureau of Standards's (NBS) Flammable Fabric Accident Case and Testing System (FFACTS) file on actual burn accidents.

Analysis of the fabric burning behavior and burn injury data indicates that, under the conditions of this test, there is less than 1 to 2 percent body area injury potential (Fig. 10) for 100 percent textured polyester, nylon, spun polyester, and flame-retarded fabrics such as FR cotton, spun FR polyester, and FR polyester in certain blends with FR rayon. A wool fabric also registered less than 2 percent body area damage. All of these fabrics self extinguished. In fact, only 3 of the 45 commercial textured polyester fabrics gave any injury at all, and that was restricted to 1 percent second-degree damage at the ignition point.

FIG. 9—*A-line garment and ignition tab.*

- *TEXTURED POLYESTER*
- *SPUN  POLYESTER*
- *NYLON*
- *FR  COTTON*
- *FR  POLYESTER*
- *FR  POLYESTER / FR RAYON*
- *WOOL*

FIG. 10—*Fabrics with less than 1 to 2 percent bodily injury.*

For the more flammable fabrics, Thermo-Man has the ability to quantify different degrees of heat transfer and injury. This is illustrated in Thermo-Man injury diagrams. In Fig. 11, a 3.2-oz untreated cotton burned 33

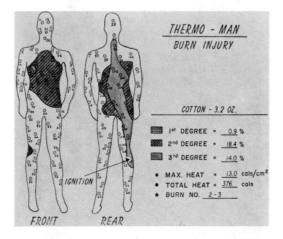

FIG. 11—*Thermo-Man burn injury, cotton (3.2 oz).*

percent of the body, with 14 percent experiencing full-thickness, third-degree burns. The dotted area denotes first-degree burns, while the lined area denotes second-degree, and solid black denotes third-degree. Figure 12 shows that a lighter-weight, 2.9-oz untreated polyester/cotton burned 29 percent total area, but only about 2 percent of the area was full-thickness injury. A-line garments of acrylic or acetate gave localized burns (largely second-degree), shown in the next two figures (Figs. 13 and 14), both covering areas of about 15 percent. The acrylic and the acetate dresses often self extinguished when flame fronts reached close-fitting garment areas.

In addition to the extent and severity of burn injury shown in these injury diagrams, our system prints out an injury-time profile. To illustrate, results for untreated 3.2-oz cotton are shown in Fig. 15. Note that it took 24 s after garment ignition for initial injury to occur, and, 6 s later, 10

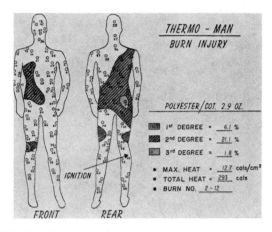

FIG. 12—*Thermo-Man burn injury, polyester/cotton (2.9 oz).*

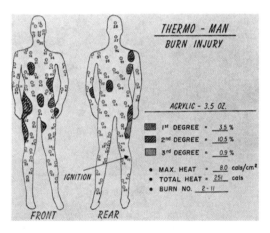

FIG. 13—*Thermo-Man burn injury, acrylic (3.5 oz).*

percent of the body was burned through the epidermis. After 40 s, sufficient heat was generated to cause full-thickness or third-degree burns, increasing up to about 15 percent of the body at test termination. In Fig. 16 for untreated 2.9-oz polyester/cotton, initial injury occurred at 12 s, and an additional 30 s were required to cover 10 percent of the body. Fifty-two seconds were required before third-degree burns occurred, which increased only to 2 percent of the body area at termination.

Following these initial experiments, some of which have been reported [8], we expanded our experimental work to include various modes and intensities of ignition sources, the effects of garment configuration and its proximity to the manikin surface, and the effects of multiple garment

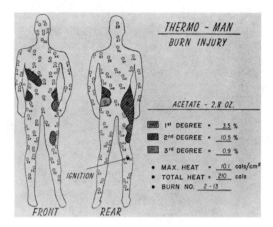

FIG. 14—*Thermo-Man burn injury, acetate (2.8 oz).*

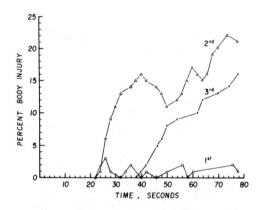

FIG. 15—*Cotton (3.2 oz) large tab behind knee.*

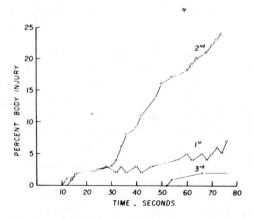

FIG. 16—*Polyester/cotton (2.9 oz) large tab behind knee.*

layers. These variables, as anticipated, complicated the apparel flammability picture to a significant degree. Results cast serious doubts on the ability of any single, small-scale test to predict the hazards of apparel products. Some specific examples are discussed.

One example of our mode of ignition work is shown in Fig. 17. In this

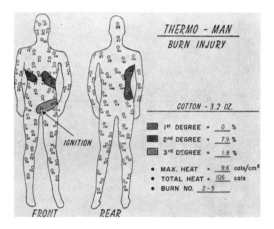

FIG. 17—*Thermo-Man burn injury, cotton (3.2 oz).*

case, we ignited the cotton A-line garment with the paper tab in the lap rather than at the hem. The flame spread was slower, giving only 10 percent area burns in 80 s. Other ignition tests (some to be discussed later) confirmed the sensitivity of untreated cotton to oxygen availability. Paper tab ignition at the hemline was selected as the standard ignition mode for A-line garments as representing the more severe case, since meltable fibers cannot shrink away from the ignition source, and air is readily accessible to both fabric surfaces.

Examples of multigarment system burns are shown in Figs. 18–20 for a

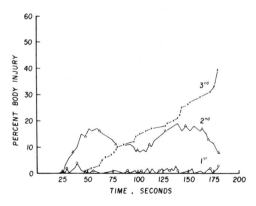

FIG. 18—*Cotton dress over polyester/cotton slip and cotton/rayon pants.*

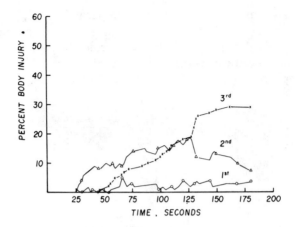

FIG. 19—*Cotton dress over nylon slip and nylon/polyester pants.*

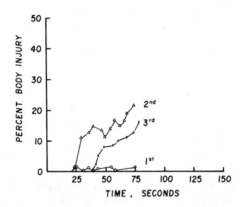

FIG. 20—*Cotton (3.2 oz) large tab behind knee.*

cotton dress over cellulosic versus thermoplastic undergarment systems. An important observation was that both undergarment systems provided a degree of protection from injury to Thermo-Man during the early stages of the garment burns. Comparing time-injury profiles of the multilayer garment burns to that of the cotton A-line garment without undergarments shown earlier (Figs. 15 and 20, the latter rescaled for comparison with Figs. 18 and 19) revealed that less second- and third-degree injury was experienced in tests with undergarments, during the first 80 s.

For example, both sets of undergarments restricted second- and third-degree injury to less than 5 percent body area at 30 s versus 11 percent area damage without undergarments. At 60 s, the nylon undergarments suppressed injury to 13 percent second- and third-degree area damage versus 20 percent with cotton undergarments as compared with 27 percent without undergarments. However, if the test garments are permitted to continue burning for an unrealistic 3 to 4 min, extensive body damage is incurred by both systems from additional fuel provided by the undergarments. Both situations of protection and aggravation exist in actual burn accidents. The FFACTS file contains examples where underclothing apparently provided protection and other examples where underclothing appeared to contribute to the extent of injury.

In order to further our understanding of the way apparel fabrics ignite and spread flame, we investigated a series of common fabrics in several standard garment configurations. In this series, we also studied the differences between ignition at hemline extremities and garment surface ignition.

We depict some of these relationships in photographic montages showing flame spread at 30-s intervals.

In Fig. 21, we show the flame spread performance of five different fabrics in dresses, with ignition at the middle of the front hemline. This is an example where visual ranking can be quite misleading. Although more apparent in color photography, these black and white picture sequences imply that the potential damage from the acrylic or polyester/cotton garments might be greater than untreated cotton (as illustrated by flaming and the "disappearance" of the white garments) during the 60 to 90-s period. However, Thermo-Man heat transfer measurements told a different story. As we can see in Table 1, the cotton garment at 80 s had caused 25 percent body area damage (second- and third-degree burns). The polyester/cotton caused 18 percent area damage, and the acrylic, 9 percent. Wool was less than 2 percent.

It is a common error among laymen to associate the heat from a flame with its luminosity in the visible portion of the spectrum; that is, the more visible the flame, the hotter. As any chemistry student knows, luminosity indicates inefficient combustion, and it is the nonluminous flame which is producing the most heat.

To illustrate these points, we have observed the thermal radiation from the flames of burning fabrics using a scanning infrared camera (Barnes Engineering Type T-6), sensitive from 2 to 20 $\mu$m. Without going into detail, this experiment showed that there is virtually no relationship between the visual luminosity and the thermal radiance of fabric flames. In fact, very little thermal radiation is emitted from the very thin flames occurring in fabric fires. So the eyes can be a poor judge of heat and hazard.

Figure 22 shows flame spread characteristics of five fabrics in men's pajama-type garments when ignited at the right pant cuff. In this case, and in contrast to front-hem ignited dresses, the visual flame spread rate was fastest for cotton. Garments of 100 percent nylon essentially failed to ignite. Typically, the acrylic is slower to get going but, like 100 percent

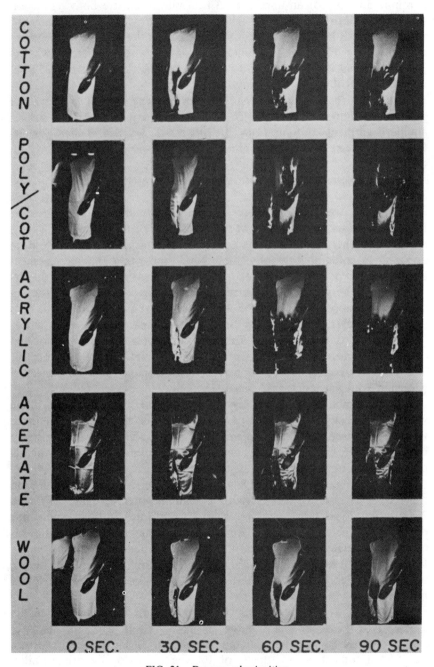

FIG. 21—*Dresses, edge ignition.*

TABLE 1—*Thermo-Man injury in dresses.*

| Composition | oz/yd$^2$ | Second- and Third-Degree Burn Area, %[a] |
|---|---|---|
| Cotton | 3.2 | 25 |
| Polyester/cotton | 2.9 | 18 |
| Acrylic | 3.5 | 9 |
| Wool | 3.5 | <2 |

[a] 80 s after ignition.

cotton and polyester/cotton, can be largely consumed if the test is continued for a longer period of time.

The next two montages (Figs. 23 and 24) show the differences between ignition at the front hemline versus ignition on the front surface of shirt-type garments. The fabric damage patterns are quite similar for all-cotton, polyester/cotton, and acrylic within a given ignition case. However, as expected, surface ignition lagged edge ignition in flame spread rate. In both ignition modes, 100 percent polyester did not propagate a flame. These findings were essentially duplicated in a pants series comparing cuff ignition with ignition on the surface at the knee.

## Flammability Test Concepts

Through hundreds of burn experiments on Thermo-Man, some of which have been described in this paper, we have gained a better understanding of garment flammability and a high level of confidence in the capability of this system to quantify injury potential from burning garments. Though not perfect, we believe Thermo-Man provides the best available reference for developing a meaningful fabric flammability test suitable for general apparel.

An overall perspective of the problem of developing a general flammability test for apparel fabrics must include the following points.

1. The hazard to be protected against must be identified and quantified as thoroughly as possible from reliable accident reports and from laboratory research.

2. We must be able to predict real-world behavior as best we can through simulation of key aspects of apparel fires.

3. Finally, we need to define intrinsic fabric properties related to the hazard and reliable laboratory-scale measuring tools which are considerably less complicated than the full-scale simulations required for the first two points. Here, the object is to provide a continuous scale along which fabrics can be ranked and meaningful pass/fail criteria selected.

As we have said, Thermo-Man enables us, for the first time, to quantitatively compare and rank garments by hazard in a laboratory situation.

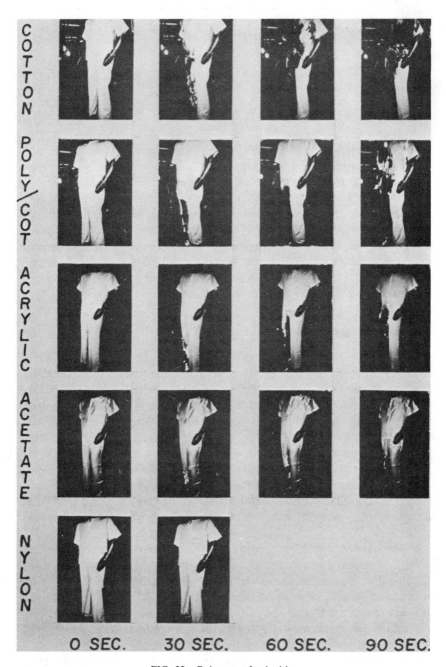

FIG. 22—*Pajamas, edge ignition.*

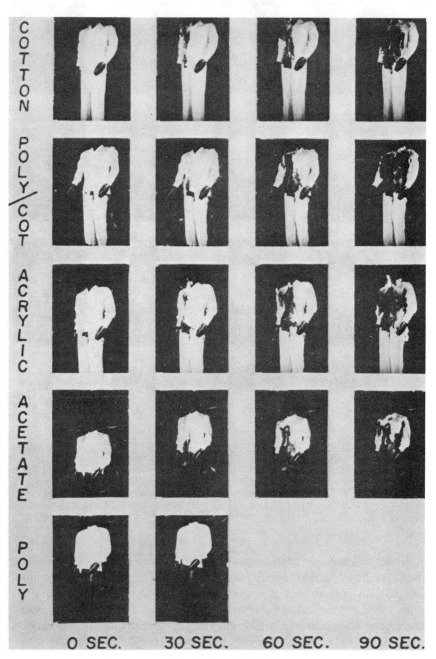

FIG. 23—*Shirts, edge ignition.*

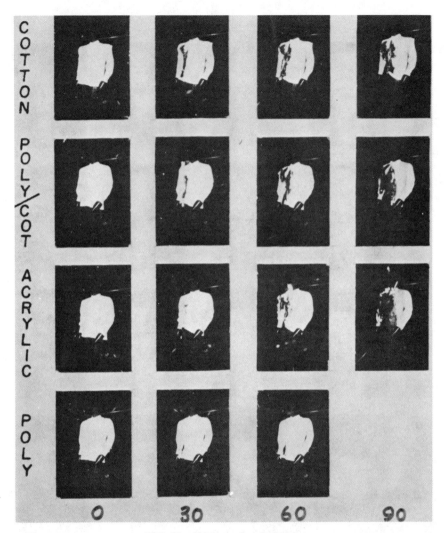

FIG. 24—*Shirts, surface ignition.*

In order to adapt our Thermo-Man results to the first two requirements listed, we have derived a parameter designated as a burn severity function, *B*. This permits a single number to be assigned which takes into account the depth and area of injury from a given Thermo-Man accident simulation. The equation for *B* is shown. Fabrics causing injuries varying in depth, as well as extent, can thus be directly compared quantitatively

$$B(A,\delta) = A[\ln(\delta/100) + 1]$$

$B$ is a function of burn area ($A$) and burn depth ($\delta$). The derivation of $B$ comprehends a linear increase in burn severity with increasing burn area so long as there is damage through the epidermis, which is assumed to average 100 $\mu$m in thickness. The severity also increases with increasing depth of burn injury but at a nonlinear decreasing rate as the depth increases.

The dependence of $B$ on depth and area is shown in Figs. 25 and 26. In Fig. 25, $B$ is plotted against burn area ($A$) at various depths of injury. This plot shows, as we have stated, that a given increment of injury depth

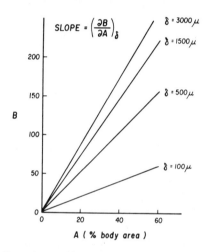

FIG. 25—*Dependence of burn severity on area at different depths.*

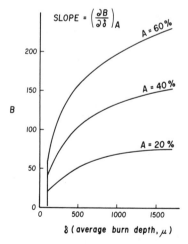

FIG. 26—*Dependence of burn severity on depth at different areas.*

has a greater effect on *B* values at shallower depths than at deeper injury depth. In Fig. 26, *B* is plotted against burn depth ($d$) at various areas of injury. If burn area is small, then *B*, the burn severity function, does not increase very rapidly as depth of injury increases. However, in the cases where burn injury areas are larger, severity increases rapidly with increasing damage depth.

We believe this system may, in fact, provide more precise burn injury information than is presently available through physicians from burned patients. There is no clear-cut, clinical method for quantitatively assessing the severity of human burn injury beyond rough estimates at the time of hospital admission. In this regard, Dr. J. A. Moylan [9] has stated "... the clinical signs we use [to estimate second- and third-degree burns] are very inaccurate. Let me emphasize that people who are experts in burn care are right only six in ten times when they say a burn is second or third [degree] .... Indeed, the proof of the pudding is, if it's second degree, it'll heal without grafting, if it's third degree, it'll require skin grafting. ... It's a functional distinction."

The Thermo-Man system, we believe, provides a valid conceptual model of burn injury, not perfect by any means, but good enough to allow us to proceed with confidence in our efforts to define the elements of a general apparel flammability test method.

Any proposed flammability test method ought to meet the requirement of predicting relative burn severity on Thermo-Man for some series of fabrics.

Table 2 shows how some current char-length tests fare versus Thermo-Man. In both the present federal children's sleepwear standard, FF 5-75, and the proposed semirestraint test, the length of char or burned fabric is intended to distinguish hazardous from nonhazardous items. An aver-

TABLE 2—*Thermo-Man versus current tests.*

| Specimen, oz | FF-5 | | Semirestraint Char Length, in. | Thermo-Man Burn Severity, *B* (A-Line Dresses) |
|---|---|---|---|---|
| | Char Length, in. | Pass/Fail | | |
| Acetate, 2.8 | 10 | F | 12 | 24 |
| Acrylic, 4.3 | 10 | F | 12 | 25 |
| Wool, 3.5 | 10 | F | 12 | 5.6 |
| 65/35 polyester/cotton, 2.9 | 10 | F | 12 | 53 |
| Cotton, 3.2 | 10 | F | 12 | 88 |
| Polyester, 7.5 | 5 | P | 6 | 0 |
| Polyester, 7.0 | 3 | P | 3 | 1.5 |
| Polyester, 6.5 | 10 | F | 10 | 2.8 |
| 90/10 polyester/silk, 7.0 | 10 | F | 12 | 5.9 |
| Polyester, 5.8 | 10 | F | 10 | 0 |

NOTE—Tests fail to consider the essential aspects of burn injury production: the burning fabric evolves heat which is transferred to the body.

age char length in excess of 7 in. represents a failure. The maximum specimen length in FF-5 is 10 in. and in the semirestraint is 12 in. In the last column are the Thermo-Man $B$ values. We see that these fabrics describe a range of injury production on Thermo-Man from a $B$ value of zero, to a $B$ value of 88, with numerous intermediate values. No such range exists in the char-length data.

It would appear that char length is a rough cut criterion with problems of regulatory overkill and shows little, if any, correlation with hazard. Numerous fabrics of very low injury potential fail the test. Further, fabrics which vary in burn severity by more than an order of magnitude are indistinguishable by the char-length tests. The problem with these tests is that they do not measure the essential aspect of burn injury production: heat which is tranferred to the body. Only when that heat is in excess of some minimum over a sufficiently large area does clinically significant injury occur.

Recognizing this, we undertook a program to characterize a variety of fabrics for their lateral heat transfer capability. We selected a fixed and controllable configuration, as shown in Fig. 27; the 8 by 9 in. specimen is held in a frame ¼ in. from a heat sensing surface. Ignition is with an 0.1-g filter paper tab. The entire apparatus is vertical. The heat sensing board, shown in Fig. 28, monitors the burning of a 4 by 5 in area in the center of the specimen. It consists of 24 blackened copper disks about ¾ in. in diameter. These disks are scanned every second by a VIDAR digital data acquisition system. All results are compared directly with identical fabrics in tests on Thermo-Man.

Figure 29 shows the excellent correlation we have obtained between heat transfer to the sensor board and the Thermo-Man burn severity for A-line dresses. The coefficient of determination, $R^2$, is 0.97, meaning that 97 percent of the variation in $B$ is accounted for by the variation in the abscissa. We will discuss the abscissa in a moment. Recognizing that some of these 13 data points represent a relatively small number of determinations on both the sensor board and Thermo-Man, we are nevertheless optimistic that the data fit will be further confirmed as more determinations are made.

The abscissa involves the average heat dose, $q$, (in calorie/square centimetre) which the 24 sensors have received for the entire duration of the sensor board burn. We divide $q$ by $t$, the time it has taken the entire sample to burn. This is, in effect, the inverse of the burning rate. Thus, the abscissa is the heat dose weighted by the rate of burning or flame spread.

We note that the ordinate involves the value of Thermo-Man $B$ at 80 s from the time of fabric ignition. The reasons for this choice have already been discussed. For fabrics which self extinguish, 80 s generally exceeds the actual burn time. However, for nonextinguishing fabrics, 80 s is con-

FIG. 27—*Fabric and ignition tab in heat transfer board.*

FIG. 28—*Heat transfer board.*

siderably shorter than the time it takes such a garment to burn to completion. For lightweight garments, it may take 3 min or so; for heavyweights, as much as 7 min may be required. Thus, there is an implicit time dependence in the Thermo-Man data as we report it. Since the burn severity depends so strongly on area involved, it is clear that, at 80 s, a smaller area will be involved with a slow-burning fabric than a rapidly burning one. Thus, we divide the area averaged heat dose that a fabric is

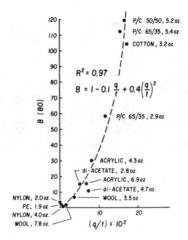

FIG. 29—*Heat transfer to Du Pont sensor board.*

capable of transferring to a vertical surface by the time it takes to burn a fixed area.

We have work currently underway to assess Thermo-Man injury potential and correlation to the sensor board for additional garment configurations, including the six specified garment types under study by the Consumer Products Safety Commission. Initial Thermo-Man data for three garment configurations in three dissimilar fiber compositions are shown in Table 3. Note that the looser-fitting nightgown and pajama garments are ranked similarly to that described earlier for A-line dresses. These data tend to confirm that various fabric compositions can develop their own unique burning characteristics in different garment types.

For the case of pajamas, an apparent chimney effect may actually promote the burning of cotton pajama pants. The proximity of the pajamas to the Thermo-Man body (in comparison with a nightgown or dress) provides greater heat transfer as long as there is adequate oxygen for combustion. On the other hand, the pajama configuration appears to slow down heat transfer and injury in the case of acrylics. Thermo-Man documentation shows that, for loose fitting garments like pajamas with ignition at the lower edge of the pants, the cotton garment is largely consumed in 80 s (Fig. 30), resulting in 63 percent area second-degree and third-degree burns, shown in Fig. 31. By comparison, the acrylic garment has been consumed only to the knee (Fig. 32) in 80 s, yielding only 4 percent area burns, shown in Fig. 33.

In the shirt configuration, we hypothesize that the closeness of fit of the cotton garment resulted in considerably less heat transfer than for pajama or nightgown-type garments. This can be explained on the basis of oxygen depletion with some heat-sink cooling effect from Thermo-Man

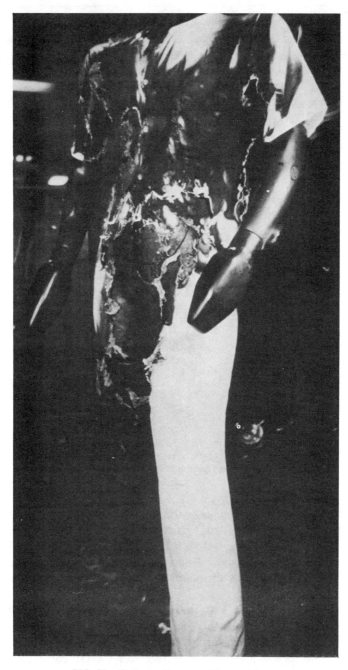

FIG. 30—*Cotton pajamas on Thermo-Man.*

TABLE 3—*Thermo-Man injury (edge ignition at lowest extremity).*

| Garment | Weight, oz/yd$^2$ | Second- and Third-Degree Injury,[a] Body Area, % | Burn Severity, Index, $B$ |
|---|---|---|---|
| Nightgowns | | | |
| cotton | 3.2 | 39.3 | 136 |
| acrylic | 3.3 | 13.2 | 51 |
| nylon | 4.0 | 0 | 0 |
| Pajamas | | | |
| cotton | 3.2 | 62.8 | 241 |
| acrylic | 6.9 | 4.1 | 16 |
| nylon | 4.0 | 0 | 0 |
| Shirts | | | |
| cotton | 3.2 | 14.8 | 59 |
| acrylic | 3.3 | 18.1 | 65 |
| nylon | 4.0 | 0 | 0 |

[a] Terminated at 80 s.

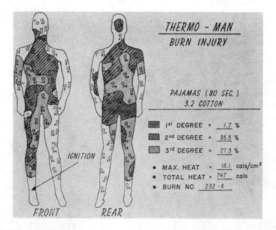

FIG. 31—*Thermo-Man burn injury, pajamas (80 s), 3.2 oz cotton.*

himself. Thermo-Man documentation for shirts was obtained 80 s after tab ignition at the lower, front shirt-tail hem. The front of the cotton shirt is well consumed (Fig. 34) with 15 percent area burns (Fig. 35). Contrary to the case for gowns and pajamas, acrylic fabric nearly duplicates (Figs. 36 and 37) cotton performance in shirts for second- and third-degree burn damage.

In still another Thermo-Man study, a series of fabrics was evaluated in dress versus shirt-type garments. Results shown in Fig. 38 basically reinforce those just discussed for gown, pajama, and shirt configurations. Figure 38 shows that oxygen depletion effect for the cotton shirt versus the dress. Also, it should be noted that, except for cotton, the injury in-

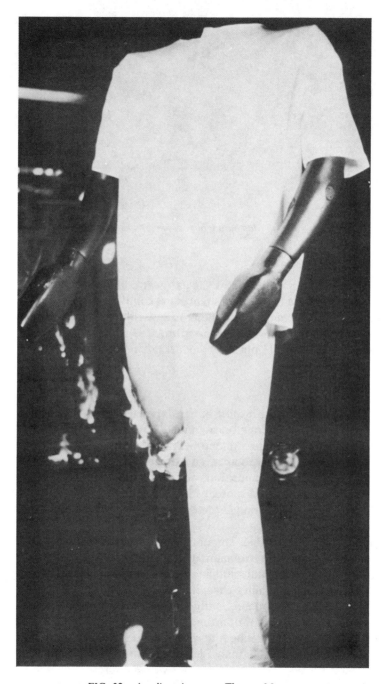

FIG. 32—*Acrylic pajamas on Thermo-Man.*

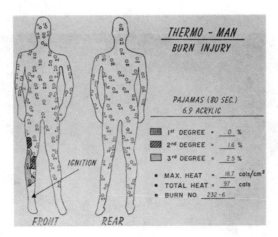

FIG. 33—*Thermo-Man burn injury, pajamas (80 s), 6.9 oz acrylic.*

dex for these fabrics increased as the garment configuration shifted from dresses to shirts, due to the shirt's closer proximity to the body.

Fabrics of 100 percent nylon failed to cause injury in any of the various garment configurations. Similar performance would be expected from garments of 100 percent polyester, modacrylic, or FR cotton.

## Conclusions

In nearly all this work, it must be reemphasized that we have elected to study the severe condition: "once ignited, how much damage resulted?" We recognize, for example, that garment style and fit can affect probability of exposure to ignition sources, an extremely important factor. Further, we know that fabric composition, construction, and condition can affect ease of ignition, once exposed. Finally, we are aware that all of these factors plus human reaction can determine how the flames spread, transfer heat, and cause damage.

Although a considerable amount of research has been done with Thermo-Man, much more remains to be done. In an industry-wide program, we are systematically studying on Thermo-Man a broad range (in both fabric and flammability characteristics) of commercial and development fabrics in basic apparel configurations to define the effect of garment type, fit, and composition on flame spread, heat transfer, and potential injury. We will relate these results to the Consumer Products Safety Commission's National Electronic Injury Surveillance System (NEISS) and FFACTS file accident data where appropriate. We will also relate the results to the various test methods under development and identify

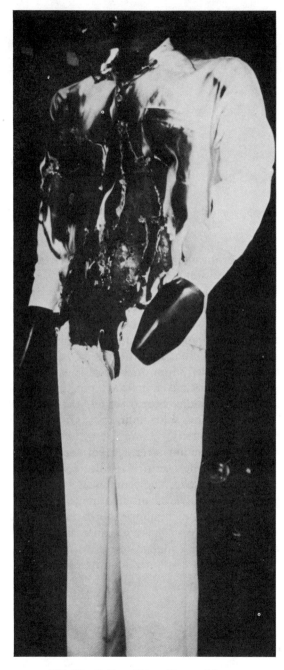

FIG. 34—*Cotton shirt on Thermo-Man.*

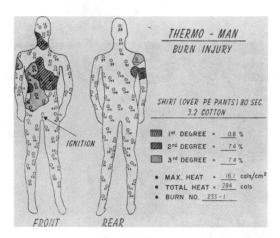

FIG. 35—*Thermo-Man burn injury, shirt (over polyester pants), 80 s, 3.2 oz cotton.*

the most appropriate, meaningful approaches. The task is very difficult but accomplishable and is underway.

It is our conviction that the data presented here have been obtained through the most advanced techniques and equipment available. Yet, we are still faced with unanswered questions and anomalies. Aside from the preliminary conclusions reported, one notable result of this program is the clear reinforcement of opinions that the subject of ignition and flammability of fabrics and the definition of hazards is extraordinarily complex. Regulations should reflect that complexity, not in a complexity of their own, but rather in strict adherence to sound scientific facts and reliable accident and economic data. With industry and government working together to determine and verify relevant data, we can develop tests and standards for use where there is a recognized need that will minimize the hazards at the least possible sacrifice to the consumer in garment performance, durability, aesthetics, and economics.

We can never expect to live in an environment that is utterly free of danger, but we certainly can minimize the hazards. To this end, many segments of our textile industry have recognized a responsibility to throw their physical, financial, and human resources into programs to provide safer products. We believe we are making progress toward that objective.

## References

[1] Chouinard, M. P., Knodel, D. C., and Arnold, H. W., *Textile Research Journal,* Vol. 43, No. 3, 1973, pp. 166–175.
[2] Henriques, F. C., Jr., *Archives of Pathology,* Vol. 43, May 1947, pp. 489–502.
[3] Stoll, A. M. and Greene, L. C., *Journal of Applied Physiology,* Vol. 14, No. 3, 1959, pp. 373–382.

FIG. 36—*Acrylic shirt on Thermo-Man.*

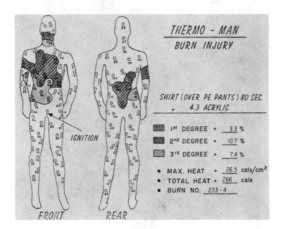

FIG. 37—*Thermo-Man burn injury, shirt (over polyester pants), 80 s, 4.3 oz acrylic.*

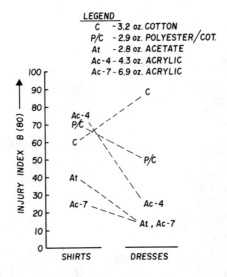

FIG. 38—*Comparison of fabrics on Thermo-Man, shirts versus dresses.*

[4] Weaver, J. A. and Stoll, A. M., *Aerospace Medicine,* Vol. 40, No. 1, 1969, pp. 24–30.
[5] Davies, J. M., Headquarters Quartermaster Research and Engineering Report T-24, April 1951.
[6] Bales, H. W., Hinshaw, J. R., and Pearse, H. E., University of Rochester Atomic Energy Report UR-438, July 1956.
[7] Moritz, A. R. and Henriques, F. C., Jr., *American Journal of Pathology,* Vol. 23, 1947, pp. 695.
[8] Bercaw, J. R. and Jordan, K. G., *Proceedings,* Eighth Annual Meeting of the Information Council on Fabric Flammability, Dec. 1974, pp. 214–239.
[9] Maylan, J. A., "How Textile Fabrics Cause and Affect Burns," University of Wisconsin, presented before the Fabric Flammability Product Liability Seminar, Dec. 1974.

# DISCUSSION

*C. F. Klein*[1] *(oral discussion)*—On the way to the aiport, my wife told me about an article in the paper; I didn't read the article myself, but she says it reported that some types of cancer are associated with fabric flame retardants. Can you comment on this?

*J. R. Bercaw (author's oral closure)*—As far as I can understand the situation, a particular chemical retardant that is used for 100 percent polyester fabrics to ensure compliance with the childrens' sleepwear standards, has recently been found to be mutagenic in the Ames mutagenicity test. According to the National Cancer Institute's (NCI) Advisory Committee on Environmental Carcinogenicity, this does not prove that such flame retardants are, in fact, carcinogenic. It means that animal tests must be run to make a determination of carcinogenicity. Animal tests have been underway at NCI for over a year and a half on this material, as well as on 80 or 90 other chemicals that they are studying. We expect results around the end of this year to tell us what the problem may be. At this point in time, it would be an error to say that this chemical is carcinogenic.

*R. E. Taylor*[2] *(oral discussion)*—In line with the work you are doing, which is excellent on burns, are you or do you propose to collect and analyze the gases which would be inhaled from those fabrics by the person who they burn and study the effects they would have, both short- and long-term, on that person?

*J. R. Bercaw (author's oral closure)*—That's outside the scope of our work. However, later in this symposium you will be hearing about some excellent work that is going on at the University of Utah and other places which will take your question into account. At the present time, they are more concerned with combustion products from materials of construction because they believe that is a greater hazard.

[1] Senior research engineer, Johnson Controls, Inc., Milwaukee, Wis. 53201.
[2] Coordinator, Codes and Standards, Republic Steel Corp., Cleveland, Ohio 44101.

*S. Martin*[3] *(written discussion)*—What temperatures are actually being measured? The paper does not seem to explain this point.

I would also like more information on the extent to which the thermal properties of the manikin have been matched to those of human skin. This is important since it is well known that any heat sink material in close proximity to a burning solid, especially thin films such as fabrics, affects the burning behavior.

*A. Z. Moss (author's written closure)*—The only temperatures measured are subsurface temperatures at a depth of 12.5 $\mu$m beneath the surface of the molded epoxy-fiberglass resin of which the manikin is made. The heat conduction properties of this material are well characterized. The test was chosen to allow convenient numerical calculation of surface heat fluxes using a one dimensional finite-difference form of the Fourier heat conduction equation. The resulting surface temperatures, though about 30 percent higher than in living tissue for similar exposure, are different enough from the source temperature that it is not necessary for the properties of the skin simulant to exactly match those of human skin. Excellent correlations have been demonstrated between the Thermo-Man and either human or animal exposure to thermal radiation. We are unaware of any effect of this device in distorting the burning behavior of fabric.

[3] Manager, Fire Research, Stanford Research Institute, Menlo Park, Calif. 94025.

*R. Friedman*[1]

# Ignition and Burning of Solids

**REFERENCE:** Friedman, R., **"Ignition and Burning of Solids,"** *Fire Standards and Safety, ASTM STP 614,* A. F. Robertson, Ed., American Society for Testing and Materials, 1977, pp. 91–111.

**ABSTRACT:** This article reviews current knowledge of four stages of fire growth: ignition, smoldering, flame spread, and burning of a fully ignited surface. Both cellulosic and thermoplastic combustibles are considered. In each case, the emphasis is that of the combustion scientist who seeks to identify controlling physicochemical mechanisms and ultimately to be able to make quantitative predictions of behavior based on understanding rather than empiricism. The present outlines of knowledge are traced, and problems particularly worthy of further research are identified. In view of the vastness of the subject, only a few highlights of the literature are discussed; 61 references are provided. Work in progress, insofar as it is known to the author, is mentioned.

**KEY WORDS:** fires, ignition, burning rate, flame propagation

This article is intended to provide a guide to current research activity on ignition and burning of solids. The emphasis is on the two most commonly encountered classes of combustible solids, namely, cellulosics (wood, paper, cardboard, cotton) and thermoplastics (polyolefins, polystyrene, polyurethanes, poly(vinyl chloride)). Research workers are particularly fond of working with purified alpha-cellulose and with poly(methyl methacrylate), the latter because it gasifies directly to the monomer, it does not drip excessively, and its properties are better defined than most other plastics.

Clearly, the flammable world surrounding us contains a far greater variety of combustible solids than this list suggests, and, furthermore, materials are used as blends or composites or contain fillers or additives of many types. However, the researcher at this time has major difficulty in explaining the combustion behavior of relatively pure cotton, polystyrene, etc. Accordingly, the research progress on "realistic" blends and composites will necessarily lag the studies on the "simpler" substances.

[1] Vice president and director of research, Factory Mutual Research Corporation, Norwood, Mass. 02062.

Even the simplest imaginable solid ignites and burns with far greater complexity than a gaseous fuel. Since the solids we are concerned with must volatilize to form gases before combustion can occur, all the complexities of gaseous combustion are present and, in addition, certain characteristics of the solid: (a) transient thermal response; (b) pyrolysis thermodynamics; (c) pyrolysis kinetics; (d) radiation properties at surface; (e) diathermanous properties; (f) thermal conductivity; (g) melting; and (h) char formation. If we recall the status of gaseous combustion research in about 1932, the first qualitative insight into the hydrogen-oxygen combustion via a branching chain reaction involving atomic hydrogen, hydroxyl, and atomic oxygen species had just emerged. It took approximately 30 years of subsequent research before rather complete quantitative understanding was reached. The methane-oxygen reaction is more complex and lagged by at least another decade. However, tremendous advances in our understanding of gaseous combustion have occurred over this 40-year period, and we now know how to obtain computer solutions for many gaseous combustion problems. The researcher can only hope that the next several decades will see comparable progress with solids.

### Piloted Ignition

In discussing ignition of a solid, as contrasted to ignition of a gas, a careful distinction must be drawn between piloted ignition and spontaneous ignition. The essence of the piloted ignition problem is that the combustion process of a solid may not sustain itself except perhaps momentarily once the source of ignition energy is withdrawn or exhausted.

Let us list the principal factors involved in the ability of a just-ignited solid to continue burning after the ignition energy flux is removed.

1. The pyrolysis gases must issue from the surface into a boundary layer at a high enough rate to permit establishment of a flame at a position far enough from the surface to avoid quenching.

2. The flame must heat the surface at an adequate rate to continue production of pyrolysis gases.

3. The rate of heat conduction (and possibly radiation) from the surface into the cold interior of the solid must not be excessive. This rate decreases with time as the solid heats up.

4. The rate of radiant heat loss from the surface outward to the colder surroundings must not be excessive.

5. The radiant input to the surface from more distant flames, or from hot gases and smoke, or from other facing hot surfaces can compensate for loss mechanisms.

In special cases, additional factors may be present which affect the energy balance at the surface, including melting and dripping, heat loss to adjacent cold inert surfaces, cooling by impinging water droplets from a sprinkler, etc.

The research on piloted and spontaneous ignition of cellulosic solids was reviewed in 1970 by Welker [1][2] and in 1972 by Kanury [2]. Kanury credits Bamford, Crank, and Malan [3] in 1946 as having initiated modern research on the problem and credits Akita [4], Simms [5], Weatherford [6], and Martin and Alvares [7] with making especially significant contributions. In very brief summary, the following has been established for piloted ignition of cellulosic solids.

Flaming will not persist in a single wood slab not subjected to supplemental external heating unless the average temperature within the slab is greater than about 320°C. The corresponding pyrolysis gas flux from the surface is greater than $2.5 \times 10^{-4}$ g/cm$^2 \cdot$s. Transient flaming can occur when the surface temperature reaches 450°C (convective heating) or 300 to 410°C (radiant heating in a quiescent ambient atmosphere).

Under ideal conditions, the time required to heat a wood slab of given thickness to a condition of sustained burning is proportional to $C\varrho x$ if the slab is thermally thin, and to $x^2/\alpha$ if the slab is thermally thick, where $x$ is slab thickness (or half thickness if appropriate), $C$ is specific heat, $\varrho$ is density, and $\alpha$ is thermal diffusivity. Thermal thickness is satisified when the slab is substantially thicker than $\lambda(T^* - T_0)/I$, where $\lambda$ is thermal conductivity of the slab, $T^*$ is "critical surface temperature for sustained ignition," and $I$ is the incident heat flux on the surface. Thus, the dividing line between thermally thin and thermally thick behavior depends on the heating rate $I$.

In a recent study, Rangaprasad et al [8] correlated radiant piloted ignition data for cotton and other thin cellulosics in a vertical position and showed that all data could be correlated in a plot of $t_i/w$ versus $aI$, where $t_i$ is time to ignition, $w$ is fabric weight per unit area, and $a$ is absorptivity for incident radiation of intensity $I$. The minimum value of $aI$ was about 1.25 W/cm$^2$.

This represents the highlights of what has been established concerning pilot ignition of cellulosic solids. The 320°C average temperature criterion, as well as the 1.25 W/cm$^2$ absorbed radiant flux criterion, while obviously useful, are purely empirical, and no attempt has been made to derive these numbers from the five fundamental factors just listed (pyrolysis gas concentration in boundary layer, heat transfer from flame to surface, internal heat transfer, radiant loss from surface, and radiant flux from surroundings to surface). Such a development is held up, not only because of mathematical complexity, but also, and primarily, because of lack of quantitative knowledge of component processes which are involved. Some areas requiring further knowledge are listed

1. Thermal conductivity of partially decomposed wood,
2. Kinetics and energetics of primary pyrolysis,

---

[2] The italic numbers in brackets refer to the list of references appended to this paper.

3. Secondary reactions of pyrolysis gases and possibly oxygen in the char layer,
4. Flame kinetics of pyrolysis gas mixture in boundary layer,
5. Convective and radiative components of heat transfer from flame to surface, and
6. Radiant emissivity of surface, which may change as char develops.

Turning now to various thermoplastics instead of wood or cellulose products, we find that very little systematic research on piloted ignition has been done. Thin thermoplastics, such as fabrics, are either ignited or locally destroyed by melting within a second or so when a small flame is applied. The limiting oxygen index test shows that many thermoplastics in a candle form can be ignited and will sustain combustion in atmospheres containing appreciably less than 21 percent oxygen, while sustained candle-like burning of a small wood stick requires as much as 28 percent oxygen. It may be speculated that this difference is primarily due to relative magnitudes of radiative loss from the burning surface, the plastic surfaces being at roughly 400 °C, while the charred wood is at perhaps 600 °C. If radiant loss varies as $T_s^4$, then a 200 °C difference in $T_s$ would mean three times as high a radiant loss rate.

While it is not possible to maintain sustained burning of a single semi-infinite slab of wood facing a cold ambient, it is easy to do so for a semi-infinite slab of many thermoplastics. Given a thick slab of poly(methyl methacrylate), the edges and corners of which are masked, a thin surface layer (a few millimetres thick) preheated to about 270 °C is a sufficient condition for piloted ignition of the surface to lead to sustained burning. Once this has been accomplished, application of a small pilot flame for a second or so will produce sustained burning. It would seem to be relatively easy to determine the energy requirement (joules/cm²) to bring a plastic slab to this condition as a function of heating rate, but the writer is not aware that this has been done.

### Spontaneous Ignition

The crucial and complex feature of the runaway chemical reaction leading to ignition must be added to the complications of piloted ignition previously discussed. Since spontaneous ignition is a more difficult phenomenon to describe in basic terms than piloted ignition, it is surprising that much more research has been done on the former than the latter. (They are both of comparable importance to the fire protection engineer.)

There are two extreme cases of spontaneous ignition of a solid: (a) the solid is a large porous mass undergoing slow oxidation which ultimately self-heats to glowing ignition and then flaming ignition (or directly to flaming ignition), or (b) the surface, subjected to external heat, emits hot pyrolysis gas which interdiffuses with ambient oxygen, and, at some point in the flow field, oxidative self-heating of a pocket of gas occurs, resulting in ignition. (There is a possible intermediate case of surface oxidation leading to ignition.)

In each case, a variety of geometries may be envisaged: for example, slab (vertical, horizontal—facing up or down), cylinder, sphere, cube, local hot spot or hot line source on surface, etc. An extensive literature exists on critical conditions for runaway heating with these various boundary conditions, featuring the transient heat conduction equation with an Arrhenius source term. (See the extensive review articles by Thomas [9] and Merzhanov and Averson [10], as well as more recent studies by Shouman et al [11], Thomas [12], and Zaturska [13].) In general, the use of these theories is restricted by lack of detailed knowledge of the chemical kinetics for practical cases of interest.

There are extensive experimental data reported for surface heating leading to spontaneous ignition of cellulosic materials (see Simms [5], Martin [7], and Akita [4]). The primary finding is that transient flaming ignition will occur once the surface reaches 500 to 650°C. The ignition is observed to initiate in the gas phase rather than at the surface. The larger the heated surface area, the more readily the ignition occurs. Simms finds that the occurrence of turbulence in the stream of gases emitted from the surface is important to the onset of ignition. He obtained photographs showing ignition to occur at the laminar-turbulent transition point in the flow field. Radiant surface heating with cold ambient air requires heating to a higher surface temperature than convective heating with hot air, as would be expected.

There are fewer studies of spontaneous ignition of plastics under precisely controlled conditions. Kashiwagi et al [14] ignited poly(butadiene acrylic acid) samples in a shock tube containing oxygen/nitrogen mixtures in a few milliseconds by using a fast-flowing oxidant atmosphere shock heated to about 2000 K. High-speed photographs showed the ignition to occur in the gas rather than at the suface. The lower the oxygen content, the further downstream the ignition occurred. The authors were not able to rule out the possibility that surface oxidation as well as gas-phase reaction was contributing to the ignition, although they favored the purely gas-phase mechanism. Ohlemiller and Summerfield [15] ignited polystyrene and epoxy samples with a laser, using fluxes of 90 to 400 W/cm², obtaining ignition in 20 to 200 ms in various oxygen/nitrogen atmospheres. Carbon black addition to polystyrene very markedly reduced the ignition delay, presumably by causing the energy to be absorbed near the surface rather than in depth. The results could be described by an energy balance equation requiring the surface to be heated to about 650°C, as long as the oxygen/nitrogen ratio is high enough to prevent a diffusion limitation to the ignition reaction.

The most sophisticated theoretical model of spontaneous ignition near an irradiated surface which takes cognizance of the observed initiation in the gas phase is that of Kashiwagi [*16*]. He developed a one-dimensional model in which fuel vapor and oxygen react in the gas phase, and heat is liberated at a rate proportional to a postulated one-step reaction

$$C_F C_O A e^{-E/RT}$$

where $C_F$ and $C_O$ are the local concentrations of fuel and oxidizer. The model permits fuel and oxygen to interdiffuse through nitrogen or other inert; all species are assumed to have the same diffusivity. The ignition is caused by incident radiation which is permitted to be absorbed in depth according to Beer's law. The solid pyrolyzes at the surface according to zero-order Arrhenius law. Various properties are taken to be constant and reradiation from the surface is neglected. The surroundings are cold.

The equations were solved numerically with a powerful computer, and the effect of variables on the solution was explored. The in-depth absorption feature of the model had a strong influence on ignition delay time at high incident fluxes. Ignition delay time was very insensitive to heat of pyrolysis. Effects of the nature of the diluent gas (nitrogen, helium, and argon) were found. Interestingly, no ignition would occur if the assumed rate parameter for the gas reaction were either too large or too small. Also, for a given gas reaction rate parameter, no ignition would occur if the pyrolysis rate parameter was either too large or too small. An explanation of the latter effect is: if pyrolysis occurs too readily, the oxygen cannot penetrate close to the surface, and the place where the proper mixture exists is too far from the surface and hence not hot enough for a runaway gas reaction. If pyrolysis occurs too reluctantly, the fuel supply is insufficient at the early stage of the induction period; later in the induction period, the oxygen in the high temperature region has been consumed, after which the fuel concentration builds up.

The foregoing is a one-dimensional model, lacking a boundary layer feature. In another study, Kashiwagi and Summerfield [*17*] developed a model containing a laminar boundary layer. In this model, the heating was convective by hot oxidant; there was no incident flux and no in-depth absorption. This differs from the previous problem in the sense that the fuel supply is governed by convective heating by the hot oxidant, and the heat from the runaway ignition reaction which occurs downstream is not coupled to the gasification process. Again a high-speed computer was used.

There is clearly a need for further refinement of models for this complex process of spontaneous ignition. There are outstanding difficulties, however. First of all, if each study led to analytical results that could be built upon by the next investigator, progress would be easier than the actual situation where each researcher must construct a more elaborate numerical program than his predecessors and then feed it to a more powerful computer.[3] Second, the assumption that the runaway reaction is a one-step process, first order in both fuel and oxygen, may prove to be an unacceptable simplification of a chain reaction. Third, the observations that the first appearance of luminosity is correlated with the point of turbulent breakup of a laminar boundary layer (for free-convective flow field) provides a challenge to the theoretician, since, leaving aside the chemistry, the fluid mechanics of laminar-turbulent transition is poorly understood.

## Smoldering

Smoldering is nonflaming combustion. It may propagate by a "front" or "wave" which involves air oxidation generally combined with pyrolysis. It may be self-sustaining, or it may require assistance from an adjacent energy source. It generally occurs in sheets of paper or cotton which are isolated by an air layer from good thermal contact with a heat sink or in bulk porous material which may be in contact with a heat sink as long as the porous material is thicker than a critical value. Flexible polyurethane foams, even if fire retardant, can smolder when in contact with a smoldering cotton fabric, and some polyurethane foams [18] will smolder in a self-sustained mode (without help from a cotton fabric).

Little published scientific research exists on the smoldering process. Key questions to be answered are:

1. What factors determine when smoldering can occur?
2. What governs the progagation rate?
3. What are the reaction products, especially the airborne ones?
4. Under what conditions will a transition from smoldering to flaming occur?

The writer understands that research is under way by T. Y. Toong and A. Moussa of Massachusetts Institute of Technology on smoldering of cotton and by M. Summerfield and T. J. Ohlemiller of Princeton and R. J. McCarter of National Bureau of Standards on smoldering of foam plastics, but no results have been published to date.

In 1957, Palmer [19] studied smoldering. He found that a minimum depth of wood sawdust or cork dust of the order of a centimetre was required for smoldering to occur in still air. The finer the dust particle size, the thinner the minimum depth. If air flowed across the surface at 2 m/s, the minimum depth was reduced to one eighth of its value in still

[3]M. Kindelan and F. A. Williams (*Combustion Science and Technology*, Vol. 10, 1975, pp. 1–19) have very recently made progress with an analytical development treating a portion of the ignition process, the temperature rise of a heated reactive solid.

air. However, in experiments with fiberboards at even higher air velocities, increase of air flow led to extinguishment. The critical velocity was 15 m/s in the propagating direction and 9 m/s in the opposite direction.

Palmer also investigated the existence of a maximum depth of a porous bed which can sustain deep smoldering. He induced smoldering at the base of a bed of wood sawdust 85 cm deep and found that propagation could still occur. The rate, when compared with that in shallower beds, was inversely proportional to bed depth.

Results taken from several investigators [19–23] for smoldering rates of various materials in still air are listed in Table 1. The geometries

TABLE 1—*Smolder propagation rates for various materials.*

| Material | Rate, cm/min | Investigator |
|---|---|---|
| Beech sawdust, 200 $\mu$m | 0.10 | Palmer |
| Coal, <104 $\mu$m | 0.10 | Cohen and Luft |
| Cocoa, <40 $\mu$m | 0.16 | Cohen and Luft |
| Lycopodium, <40 $\mu$m | 0.19 | Cohen and Luft |
| Fiberboard, 0.27 g/cm$^3$ | 0.20 | Palmer |
| Sawdust, undried, 76 to 152 $\mu$m | 0.24 | Cohen and Luft |
| Cigarette, between puffs | 0.30 | Gugan |
| Paper roll, 0.66 cm dia, 40°C | 0.30 | Kinbara et al |
| Paper roll, 0.4 cm dia, 40°C | 0.44 | Kinbara et al |
| Cardboard 1 cm wide, 0.076 cm thick, 60°C | 0.60 | Kinbara et al |
| Cotton, single yarn, 0.028 cm dia, 0.7 g/cm$^3$ | 5.4 | Our laboratories |
| Cotton string, 0.045 cm dia, 0.22 g/cm$^3$ | 10.8 | Our laboratories |

varied from case to case. As may be seen, smoldering propagates through a bed of porous material at a rate of 0.1 to 0.3 cm/min and propagates considerably faster along thin strands of material surrounded by air. An inverse proportionality has been noted by Heskestad (private communication, Factory Mutual Research Corporation Laboratories) between rate and the product of strand diameter and strand density.

Palmer found that smoldering rate within a bed increased sharply with increased air velocity over the bed when the velocity was in the direction of smolder propagation. He found relatively small effects of bed porosity, moisture content, and particle size.

The rate of smoldering of vertical strands is, of course, greater in the upward than in the downward direction, because of convection.

Kinbara et al [21] show a pronounced increase in smoldering rate with increasing ambient temperature. (In one case, an increase of ambient from 40 to 150°C caused a 60 percent increase in rate.)

In the writer's opinion, there is no adequate quantitative theoretical explanation of these trends.

Common observation reveals that smoldering is generally accompanied by generation of smoke which appears to be an aerosol of condensed high molecular weight species formed by pyrolysis in or near the smoldering zone. Transition from smoldering to flaming combustion may be marked by reduction of smoke formation per unit of heat release for the obvious reason that the smoke is combustible. It must be noted, however, that, when smoldering is occurring deep within a porous bed, no visible products may emerge until the smoldering zone reaches the surface because of self-absorption of products within the bed.

If the geometry of the smoldering substance controls the availability of oxygen to the smoldering zone, and this, in turn, influences the temperature within the zone, and, further, if this temperature is influenced by other factors, such as transverse heat loss or convection through the bed, would we not expect the reaction products to depend in some way on this temperature? The answer is not known, and this question must be kept in mind in trying to generalize any data.

Of course, we know that when we puff on a cigarette or blow on a smoldering object the glow intensity increases. With a cigarette, Egerton et al [22] measured a peak temperature of 820°C before it was puffed, with a thermocouple, and 1100 to 1200°C for hot spots during the puff, with a photographic technique. Palmer [19], using an optical pyrometer with smoldering cork dust, reported 760 to 780°C in still air and 900 to 930°C at 1 m/s air flow.

We are not aware of any data on the fraction of available combustion energy released during smoldering or on complete chemical analysis of smoldering reaction products. (Of course, much information is available on pyrolysis reaction products in inert atmospheres.)

Reference should be made to the National Bureau of Standards test for smoke production. This employs a closed cabinet in which a specimen with exposed surface area 6.5 by 6.5 cm is irradiated at a flux level of 2.5 W/cm². The test may be performed under either flaming or non-flaming conditions. Reduction of light transmission vertically across the chamber, because of smoke generated, is measured as a function of time. Such data are reported by Gaskill [24] and by Hilado [25]. Smoke generated during a test under nonflaming conditions might be due to smoldering (as defined herein), or it might be due to pyrolysis with no oxidation whatsoever taking place. Thus, the data available from such tests under nonflaming or so-called smoldering conditions may only be vaguely related to smoke production under natural smoldering conditions, that is, without a continuous external heat source.

As for factors which may influence the transition from smoldering to flaming, there is very little quantitative information. In Palmer's [19] experiments on effect of air flow on smoldering rates of fiberboards, he never observed a transition to flaming, even though he explored a wide range of air flows, both in the propagation direction and in the opposite direction. However, he notes that "combustion with flame could be produced with board strips if the strips were allowed to smolder in still air and accumulate a carbonized residue before the application of an air draught." On the basis of the earlier discussion of spontaneous ignition, we suppose that this procedure in some way brought pyrolysis gases at around 600°C in contact with air.

## Flame Spread

The rate of advance of a flame across the surface of a combustible solid is readily measured. It is found to be affected by many variables.

*Physical and geometrical parameters*
   Orientation of surface
   Direction of propagation
   Thickness of specimen
   Roughness or presence of sharp edges
   Initial temperature
   Pressure
   Velocity of atmosphere
   External radiant flux
*Chemical parameters*
   Composition of solid
   Composition of atmosphere

Experimentally measured effects of these variables have been summarized by Friedman [26] and by Magee and McAlevy [27]. The bulk of these studies are concerned with horizontal or downward propagation, even though upward spread occurs several orders of magnitude faster and appears to be controlled by different processes.

Since the reviews [26,27], further studies seeking to pinpoint the details of the horizontal or downward spread mechanism have been carried out by Hirano et al [28–30], Sibulkin et al [31–33], Moussa et al [34], and Fernandez-Pello and Williams [35]. The principal controversies concern the contributions of the various modes of heat transfer at the leading edge of the flame which may be different for thick versus thin materials, cellulosic versus thermoplastic materials, etc. The details of the gas motion near the leading edge are involved.

Theories of this type of propagation were reviewed by Sirignano [36] in 1972. A more current review is now being prepared by F. A. Williams for the forthcoming Sixteenth (International) Combustion Symposium.

Some novel data on the effect of preheat on flame spread across thermoplastics have been obtained by Perrins and Pettett [37]. Kashiwagi [38] has analyzed effect of radiative flux on flame spread over porous material.

The much more rapid upward spread process has been studied very little. Hansen and Sibulkin [39] made a preliminary study of spread up a small thermoplastic wall. Orloff et al [40] made a much more detailed study of turbulent spread up thick vertical polymethyl methacrylate (PMMA) walls 157 cm high. They used cooled side walls to ensure truly two-dimensional spread. The velocity of the ascending fuel pyrolysis front increased from an initial value of about 0.05 cm/s to a final value of 0.6 cm/s, still increasing when the front reached the top. Measurements of flame height were made as the propagation occurred, and the excess of the flame height over the pyrolysis front height increased continuously. From this measured flame height, combined with radiative flame heat transfer measurements and known transient thermal properties of the PMMA, values were calculated for rate of flame spread, which were in good agreement with the measured values. These results show that the spread rate is controlled by the largely radiative heat transfer from the upper part of the turbulent flame to the portion of the wall above the pyrolysis front. A further unpublished experiment by Orloff and co-workers with a taller PMMA wall, 356 cm high, shows that this spread process is still accelerating, even up to this height.

The turbulent upward spread over fabrics should differ from the foregoing behavior in that, as the flame climbs, the lower portion of the fabric burns out, limiting the ultimate size of the flame and hence the spread rate.

Markstein and de Ris [41] have studied this with cotton fabrics held vertically and at various angles of inclination, again with side walls. With fabrics up to 150 cm long, the flame spread was found to continually accelerate, acquiring a turbulent character after a brief laminar period. An analysis showed that an asymptotic rate would have been reached eventually if the fabrics had been larger. The asymptotic rate varied from 20 to 45 cm/s, depending on angle of inclination and relative humidity. Without side walls, the transverse air flow had a profound effect, shortening the rising flames and thereby reducing the distance over which they can preheat the fabric. Under this condition, the spread rate is lower than with side walls and attains a constant value rather early after an initial accelerating phase.

**Burning Rate**

For an ignited thick solid object of finite size, a more or less steady-state burning period may be expected after the flame has spread to involve all the exposed surface. The burning rate during this steady period is of considerable interest from a fire safety viewpoint because: (a) the intensity of the fire strongly influences the probability of secondary ignition of nearby objects; (*b*) the fire intensity determines the rate of buildup of smoke and toxic gases throughout the structure containing the fire; and (*c*) any given fire suppression technique (hand-held extinguisher, automatic sprinkler, etc.) becomes ineffective when the fire intensity is greater than some critical value. Accordingly, research to predict the steady burning rate of a fully involved combustible object in terms of the physicochemical properties of the combustible is important.

It is known that, if the fire is burning in a compartment with limited ventilation, or the fuel is in a geometry like a closely packed wood crib with limited air access to the interior, the burning rate will be governed by the ventilation rate which, in turn, is generally buoyancy controlled. The research on critical conditions for ventilation control is outside the scope of this review. We shall limit ourselves here to prediction of burning rate under conditions of unrestricted air supply.

The problem can be approached scientifically in six steps of progressively increasing complexity:

1. Burning rate of an evaporating liquid droplet (no convection or radiation),
2. Burning rate of an evaporating liquid surface (laminar flow only, no radiation),
3. Introduction of turbulent flow,
4. Introduction of radiative along with convective heat transfer,
5. Substitution of pyrolyzing solid for evaporating liquid,
6. Consideration of more complex geometries (that is, ceiling, pair of opposing walls, corner, inside of duct, etc.).

In all these cases, one is encouraged by results to date to make the assumption that the fire is treatable as a classical diffusion flame [42]; that is, the chemical kinetics of both the gaseous combustion reaction and the surface pyrolysis reaction (for solids) may be ignored to a first approximation. There are two coupled rate-controlling processes, the interdiffusion rate of fuel vapors and oxygen, which determines the flame size and location, and the feedback rate of combustion-released energy to the fuel surface. The latter depends in part on the temperature difference between the flame and the surface; and the surface temperature of a pyrolyzing solid depends on the pyrolysis kinetics. However, this latter dependence is generally very weak (a large change in pyrolysis rate can occur with only a small change of surface temperature). Hence, a theory assuming a constant surface temperature may give a good approximation of the process.

Much published work on this general problem is available in the classic combustion literature [42] and need not be reviewed here. The writer, in a previous review [43], has discussed some of the factors involved. In short summary, as long as we restrict our attention to sufficiently small flames so that the radiative flux is small relative to convection, excellent agreement is obtained between measured values of burning rate of simple fuels and theoretical predictions. The most important properties of the combustible which are needed for these predictions are the energy required to vaporize or pyrolyze it, $L$, the stoichiometric mass fuel-air ratio, $r$, and the heat of combustion per unit mass of fuel vapor, $\Delta H$. These are involved through a dimensionless mass-transfer number, the Spalding $B$ number, defined as

$$B = \frac{r\Delta H - C_g(T_s - T_a)}{L + C_s(T_s - T_i)} \tag{1}$$

where

$C_g$ and $C_s$ = specific heats of gas and solid (or liquid),
$\quad T_s$ = surface temperature,
$\quad T_a$ = ambient air temperature, and
$\quad T_i$ = interior temperature of the solid (or liquid).

The numerator is approximately the energy released by combustion per unit mass of air reacting, while the denominator is the energy required to gasify unit mass of fuel. The $B$ number, a thermodynamic parameter, ranges from about three to eight for most common liquids and is much smaller for pyrolyzing solids. It has not been accurately determined for many solids, but is believed to be about 1.3 to 1.7 for PMMA and somewhat lower for other solids [43–45]. It may not be a purely thermodynamic parameter, as the pyrolysis mechanism and, hence, the endothermicity of a given substance may depend on the surface heating rate. Furthermore, it has been proposed [46,47] that surface oxidation is an important part of the burning process, and this, if true, would influence the energy balance in a manner not considered in the theory. Finally, charring materials like wood burn with an initially high rate which reduces as the insulating char layer builds up, and this complication, as well as others associated with grain structure, cracking, etc., make it difficult to apply precise quantitative treatment to wood or related materials.

When the $B$ number mass-transfer approach is used to predict burning rate $\dot{m}''$ (g/cm²·s), an expression of the form

$$\dot{m}'' = \frac{\lambda_g/C_g}{\delta} \ln(B + 1) \tag{2}$$

may be derived, where $\delta$ is the thickness of a boundary layer between the flame and the surface (governed by fluid-mechanical effects) and $\lambda_g$ is the thermal conductivity of the gas. Since, in the theoretical development, the assumption is made that the Lewis number of the gas is unity

$$\frac{D_g C_g \varrho_g}{\lambda_g} = 1 \tag{3}$$

then Eq 2 may be written in the alternate form

$$\dot{m}'' = \frac{D_g \varrho_g}{\delta} \ln (B + 1) \tag{4}$$

where $D_g$ and $\varrho_g$ are the diffusion coefficient and density of the gas mixture.

For a droplet in a stagnant atmosphere, the boundary layer thickness $\delta$ may be shown theoretically to be equal to the radius of the droplet. Theory and experiment agree well for burning droplets of various liquids.

For a solid object of characteristic length $y$ burning with a boundary layer $\delta$ due to buoyancy induced by gravitational acceleration $g$ and possibly also an imposed flow velocity $V$, $\delta$ is given by a dimensionless relation of the form

$$\frac{\delta}{y} = f(\text{Gr}, \text{Re}) \tag{5}$$

where Gr is Grashof number (the square of the ratio of the buoyancy force acting on an element of hot gas to the viscous force acting to suppress flow, and Re is Reynolds number (the ratio of inertial to viscous force acting on an element of fluid)

$$\text{Gr} = \frac{g y^3}{v_g^2} \times \frac{\Delta T}{T}; \quad \text{Re} = \frac{V y}{v_g} \tag{6}$$

where $v_g$ is kinematic viscosity of the gas, and $\Delta T$ is the temperature difference between hot and cold region.

Specific forms of Eq 5 have been developed for various geometries. The most detailed analysis has been made for laminar burning at a vertical surface with a natural convection flow field. Kim et al [48] have analyzed this problem in detail, generalizing a previous treatment by Kosdon et al [49]. They find that, in addition to the primary dependence of burning rate on $B$ and Gr, there is also minor dependence on the surface temperature, the stoichiometric ratio (in addition to its effect on $B$), and a transport property parameter. For surfaces no taller than about 10 cm, flames remain laminar, and this theory gives results in good agreement with experiment.

Unfortunately, the foregoing theoretical approach becomes grossly inaccurate when one attempts to apply it to somewhat larger fires with turbulence and especially with large radiant flux. For example, Magee and Reitz [50] found that the steady burning rate of a vertical slab of PMMA 35 cm high would triple in magnitude when an external radiant flux of only 1.7 W/cm² was imposed. Radiant fluxes up to an order of magnitude larger than this have been measured near realistic fires. As another demonstration of the role of radiation, Modak and Croce (Factory Mutual Research Corporation (FMRC), not yet published) measured burning rate of a series of deep PMMA pools of different sizes:

| Pool Area, cm² | Burning Rate After 30 min, g/m²·s |
|---|---|
| 232 | 6.1 |
| 929 | 10.5 |
| 3716 | 18.5 |

Radiation measurements and analysis confirmed that this increase of burning rate is due to the increased radiative flux from the larger, more optically thick flames.

Orloff et al [40] have made detailed measurements of local steady burning rates and local flame-generated radiant energy fluxes at various heights on a vertical PMMA slab 157 cm high. Key results are:

| | Heat Flux, Flame to Surface, W/cm² | | Burning Rate, g/m²·s |
|---|---|---|---|
| Height, cm | Radiative | Convective | |
| 38 | 1.35 | 0.65 | 7.1 |
| 51 | 1.47 | 0.61 | 7.6 |
| 76 | 1.79 | 0.41 | 8.3 |
| 102 | 1.93 | 0.49 | 9.5 |
| 127 | 2.11 | 0.51 | 10.6 |
| 153 | 2.15 | 0.55 | 11.5 |

The analysis included consideration of radiant absorption by gaseous species between the flame and the surface and allowed for radiant loss from the surface outward. These results clearly show the dominant influence of radiation on fires of this size range. A recent extension of this study to a taller PMMA wall 356 cm high showed that the burning rate continued to increase with height, being 20 g/m²·s at the upper point. It is estimated that about 90 percent of the heat flux impinging on this point was radiative.

These types of results argue that detailed research on the radiative aspects of solid combustible burning is urgently needed. The problem may be divided into two parts: (a) the radiative emission from the flame as influenced by size, chemical nature, and possibly turbulence and (b) the response of a given burning solid surface to an imposed radiant flux level.

Present knowledge of flame radiation is grossly inadequate for quantitative prediction. Both soot particles and gaseous water ($H_2O$) and carbon dioxide ($CO_2$) are known to contribute significantly. On the basis of a 1400 K measured temperature (Schmidt method) for turbulent PMMA flames (FMRC, unpublished), blackbody radiation would be 22 W/cm². This is an order of magnitude larger than the earlier-mentioned radiative flux on a PMMA wall at the 1.5 m height. Thus, something far more sophisticated than an optically thick assumption is needed.

Felske and Tien [51] have recently explored a means of calculating emissivity of luminous flames, and Tien is continuing this work. Sibulkin [52] ·has made estimates of effect of flame size on radiation. Markstein [53] has meassured emission and absorption with an array of up to ten laminar diffusion flames with various gaseous fuels. The most sooty flame (1, 3-butadiene) had five times the radiance of methane and 1.7 times the radiance of propane on the basis of a single flame. The variation of transmittance with number of flames showed that, of the nine fuels investigated, the three sootiest could be represented by a grey-gas model. The other, less sooty, fuels required a model using a sum of two weighted grey-gas terms. It is suggested that these terms may be associated with soot and molecular band radiation, respectively.

In regard to the response of a burning surface when stimulated by an external radiant flux $\dot{q}_R{}''$, Tewarson and Pion [54] have obtained results which, as a first approximation, may be represented by a simple relationship

$$\dot{m}'' = \dot{m}_0'' + \dot{q}_R''/L \qquad (7)$$

where $L$ is the energy required to pyrolyze unit mass of the solid (including the energy needed to bring it from its initial state to the pyrolysis condition). By measuring $\dot{m}$ versus $\dot{q}_R''$ for a variety of solids (10 cm diameter) and taking the slope to be $1/L$, they obtain values of $L$ such as:

| Combustible | L (cal/g) |
| --- | --- |
| Polyurethane foams (various) | 291 to 364 |
| PMMA | 386 |
| Polystyrene | 420 |

| | |
|---|---|
| Polyethylene | 555 |
| Polyoxymethylene | 580 |
| Fire-retarded polyisocyanurate foam | 877 |

The value for PMMA is in excellent agreement with independently measured values; in the absence of any other method, the foregoing laboratory-scale procedure is recommended to get a general idea of how the burning rates of various solids respond to radiative flux. The foregoing list shows that some solids are three times as resistive as others. (Roberts [55] has found that burning rate of polyurethane foam increases more rapidly with radiative flux than does wood, but he attributed the difference to pyrolysis kinetics, while Magee [50] and Tewarson [54] propose that it is simply governed by the energy balance at the surface.)

From a fundamental viewpoint, the situation is more complex than Eq 7 suggests, since, as the burning rate increases with increasing external radiant flux, other factors may change, such as the convective heat transfer rate, the flame radiation from the larger flame, the rate of heat loss from the sample, etc. Analysis of these factors is now under way at FMRC.

Finally, the role of turbulence in burning rate must be discussed. The most obvious effect is on the flame size and shape, which must be known in order to predict the radiative flux which impinges on a target with a given view factor of the source flame. Thomas [56] and Steward [57] review and provide correlations of the size of buoyancy-dominated flames. Steward's model predicts that the visible flame tip corresponds to the height at which 400 percent excess air has been entrained. This demonstrates the important effect of turbulence on the flame structure.

Another effect of turbulence is to augment the heat transfer from the flame to an adjacent pyrolyzing surface. de Ris and Orloff [58] have investigated the effect of surface orientation on burning rate, including floor, wall, ceiling and intermediate orientations. Results show large dependence of burning rate on orientation, ascribed to gravitational generation of turbulent kinetic energy (maximum for an upward-facing surface), which modifies the flame thickness and the resultant radiative flux.

A promising approach to providing a more quantitative description of the flow field in a fire is offered by the application of recently developed numerical turbulence models (see Lockwood and Naguib [59]). At FMRC, an attempt is now being made to extend this type of model to include radiation.

The radiative properties of a turbulent diffusion flame may be different from those of a laminar diffusion flame of the same size. Markstein [60] has studied arrays of laminar and turbulent ethane flames and has found important differences. For example, the radiance extrapolated to an in-on. That is, the difficulty in predicting these conditions make predicting smoke movement more difficult.

There are various means for controlling smoke movement. "Smoke control design" can be defined as a way to find a means or system of controlling the smoke which is well suited to each individual building and then to derive the optimum quantitative conditions required for those means. Smoke control design, up to the present, is neither widespread in its concept nor systematized as an engineering system. From now on, it may be necessary to arrange and establish a theoretical system for design with the following three items as its skeleton.

1. Data that provide hypothetical conditions for design in calculation.
2. Calculation methods that allow quantitative evaluation of the smoke control effect.
3. Methods of introducing the calculation results into the control systems.

Item 3 is necessary to assess the results obtained from Items 1 and 2 in light of the reliability of Items 1 and 2. In the case of Item 3, it will be necessary to arrange methods and standards for assessing reliability and establishing safety factors. Here, Items 1 and 2 are only used for detailed description because Item 3 is a future rather than a pressing problem compared with the other two.

## Hypothetical Conditions in Smoke Control Design

It is almost impossible to predict when, where, and under what conditions a fire will occur. However, it may be possible to limit the scope of the probable conditions to some extent by studying the probabilities derived through analyzing data on statistics of fires and meteorology, actual conditions of various buildings, and other factors concerned. As for the conditions which have a great effect on smoke movement in buildings, that is,

1. Times of fire outbreaks.
2. Meteorological conditions (direction and velocity of wind, air temperature, wind pressure coefficient).
3. Location of fire outbreaks (floor and compartment of origin).
4. Flow resistances of doors, windows, ducts, and shafts (flow coefficients, ratio of broken windows and opened doors, etc.).
5. Temperature distribution in buildings (fire compartment, general compartments, staircases, ducts, and shafts).

it may be adequate at the present time to establish a standard for each of these conditions for input data so that the designed system is always on the safe side.

For the sake of convenience in designing the control systems in Japan, we offered the following tentative proposals for the five conditions just described. However, many of these are not supported statistically or experimentally, and there remains much future study to be done, especially the conditions of open doors and windows at the time of fire outbreaks.

### Times of Fire Outbreaks

Wintertime (buildings being heated) and summertime (buildings being air conditioned), during which a great temperature difference exists between the external temperature and the air within buildings should be chosen as a design condition.

### Meteorological Conditions

They should be determined as follows on the basis of the meteorological statistics in each particular region during winter and summer.

#### Air Temperature

Winter    Minimum average monthly values of the daily lowest temperatures (example: $-1.5°C$, January, Tokyo)

Summer    Maximum average monthly values of the daily highest temperatures (example: $30.7°C$, August, Tokyo)

*Wind Velocity*—Wind velocity $V_0$, up to that which encompasses 95 percent of accumulated wind velocity frequency observed at the observatory in each area during winter and summer, is set as a standard, and wind velocity $V$, which corresponds to the arbitrary height $h$ of a building, is determined as follows according to the site of each building.

For high-rise buildings built on broad sites

$$V = k \times V_0 (h \leqq h_0)$$
$$V = k \times V_0 (h/h_0)^{1/k} (h > h_0)$$

For buildings built within urban areas

$$V = k \times V_0 (h \leqq h_0),$$
$$V = k \times V_0 (h/h_0)^{1/k} (h > h_0)$$

where

$k$ = revising factor for wind velocity and

$h_0$ = height from the ground to the place where the anemometer of each observatory is located.

*Wind Pressure Coefficient and Wind Direction*—The standard wind direction is taken so that it is normal to the principal openings or (windows) and hence causes positive wind pressure on them. The wind pressure

[42] Williams, F. A., *Combustion Theory,* Addison-Wesley, Reading, Mass., 1965.

[43] Friedman, R., *Journal of Fire and Flammability,* Vol. 2, 1971, pp. 240–256.

[44] Holve, D. J. and Sawyer, R. F., "Diffusion Controlled Combustion of Polymers," Fifteenth, Symposium (International) on Combustion, Combustion Institute, Pittsburgh, Pa., 1975, pp. 351–361.

[45] Kanury, A. M., "Modeling of Pool Fires With a Variety of Polymers," Fifteenth Symposium (International) on Combustion, Combustion Institute, Pittsburgh, Pa., 1975, pp. 193–202.

[46] Burge, S. J. and Tipper, C. F. H., *Combustion and Flame,* Vol. 13, 1969, pp. 495–505.

[47] Stuetz, D. E., DiEdwardo, A. H., Zitomer, F., and Barnes, B. P., *Journal of Polymer Science,* Vol. 13, 1975, pp. 585–621.

[48] Kim, J. S., de Ris, J., and Kroesser, F. W., "Laminar Free-Convective Burning of Fuel Surfaces," Thirteenth Symposium (International) on Combustion, Combustion Institute, Pittsburgh, Pa., 1971, pp. 949–961.

[49] Kosdon, F. J., Williams, F. A., and Buman, C., "Combustion of Vertical Cellulose Cylinders in Air," Twelfth Symposium (International) on Combustion, Combustion Institute, Pittsburgh, Pa., 1969, pp. 253–264.

[50] Magee, R. S. and Reitz, R. D., "Extinguishment of Radiation Augmented Plastic Fires by Water Sprays," Fifteenth Symposium (International) on Combustion, Combustion Institute, Pittsburgh, Pa., 1975, pp. 337–347.

[51] Felske, J. D. and Tien, C. L., *Combustion Science and Technology,* Vol. 7, 1973, pp. 25–31.

[52] Sibulkin, M., *Combustion Science and Technology,* Vol. 7, 1973, pp. 25–31.

[53] Markstein, G. H., "Radiative Energy Transfer From Gaseous Diffusion Flames," Fifteenth Symposium (International) on Combustion, Combustion Institute, Pittsburgh, Pa., 1975, pp. 1285–1294.

[54] Tewarson, A. and Pion, R. F., "Flammability of Plastics: I. Burning Intensity," *Combustion and Flame,* in press.

[55] Roberts, A. F., *Fire Technology,* Vol. 7, 1971, pp. 189–200.

[56] Thomas, P. H., "The Size of Flames From Natural Fires," Ninth Symposium (International) on Combustion, Combustion Institute, Pittsburgh, Pa., 1963, pp. 844–859.

[57] Steward, F. R., *Combustion Science and Technology,* Vol. 2, 1970, pp. 203–212.

[58] de Ris, J. and Orloff, L., "The Role of Buoyancy Direction and Radiation in Turbulent Diffusion Flames on Surfaces," Fifteenth Symposium (International) on Combustion, Combustion Institute, Pittsburgh, Pa., 1975, pp. 175–182.

[59] Lockwood, F. C. and Naguib, A. S., *Combustion and Flame,* Vol. 24, 1975, pp. 109–124.

[60] Markstein, G. H., "Radiative Energy Transfer From Turbulent Flames," ASME Paper 75-HT-7, American Society of Mechanical Engineers, Aug. 1975.

[61] Kung, H-C., "The Burning of Vertical Wooden Slabs," Fifteenth Symposium (International) on Combustion, Combustion Institute, Pittsburgh, Pa., 1975, pp. 243–253.

# DISCUSSION

*P. H. Thomas*[1] *(oral discussion)*—I'd like to congratulate Dr. Friedman and recommend careful study of his paper. It is a very exhaustive and thorough survey. I'd like to add something to the conclusions of it. I wholly endorse what the speaker says about the lack of appreciation of the importance of thermal radiation in fires, and it has a very direct consequence for the practical business of making standards. It may well be

[1] Fire Research Station, Boreham-Herts, United Kingdom WD6 2BL.

much better in some situations for the standards to be written in terms of thermal flux, not temperature. There are standards which have been written in terms of temperature where one has to take action, as we did in International Council for Building, Research, Studies and Documentation (CIB) last week, to try to improve standardization by standardizing the uniformity of flux during the conduct of fire endurance tests. The problem of radiation is a very important one that affects standards as well as scientific research.

*R. J. McCarter*[2] (*written discussion*)—I have some comments on smoldering combustion that are prompted by recent work, to this effect:

1. Cellulosics are the primary fuel of smolder, at least in regard to apparent life-hazards.
2. Cellulosics do not have an intrinsic smolder tendency.

Regarding the latter, cellulosics have a wide range of smolder behavior, dependent upon acquired inorganic contaminants. Pure cellulose (for example, "ashless" filter paper or USP absorbent cotton) does not smolder. Thus, the answers to your first listed questions, the when and how of smolder, would be supplied for cellulosics primarily by incisive chemical analysis. The last two questions, from a life-safety viewpoint, would appear of secondary interest, as potentially lethal amounts of carbon monoxide (CO) are released in either event.

*R. Friedman* (*author's written closure*)—These are very interesting comments. Thank you.

*R. G. Gann*[3] (*written discussion*)—I would like to add a brief description of our recent studies at Naval Research Laboratory to Dr. Friedman's review of the nearly neglected realms of smoldering combustion. Using a 2.7-cm inside diameter flow tube, I. Cheng and I have investigated the smoldering of coconut shell charcoal. Both gas velocity upward through the fuel (7 to 28 cm/s) and oxygen content of the entering gas (7 to 20 percent) were varied. As might be expected, we found that the combustion temperature and mass burning rate of the fuel decreased as the oxygen fraction or gas velocity was decreased. However, the CO/carbon dioxide ($CO_2$) ratio in the effluent gas also decreased. Since the fuel burned from the bottom of the bed, we attributed this behavior of the $CO/CO_2$ ratio to a temperature-dependent, heterogeneous reduction of $CO_2$ in the downstream (that is, physically upper) layer of charcoal. A more detailed account of this work will be published within a year.

*R. Friedman* (*author's written closure*)—Thank you. We will look forward to this published account.

[2]Chemical engineer, Center for Fire Research, National Bureau of Standards, Washington, D.C. 20234.
[3]Research chemist, Center for Fire Research, National Bureau of Standards, Washington, D.C. 20234.

That is to say, this method intends to solve the movement of smoke and air in a building as a whole by computing the quantity of smoke and air flowing through each opening and flowpath, on the assumption that the temperature is uniform in each compartment except the vertical shafts, and without giving special thought to the localized flows in the building, for example, inside a certain room.

When calculation is made of smoke movement and its control, the pressure and flowrate in each part of a building must be computed, and, as the case demands, it also becomes necessary to carry out calculations for the concentration of smoke and gases, and heat in order to obtain the temperatures in each part of the building.

### Calculation of Pressure and Flow Rate

Under the same conditions of temperature and pressure, there is no remarkable difference in fluid properties between smoke and air. Consequently, it is convenient to regard smoke in the same light as air and calculate only the flow of air, and, as the need arises, to consider the concentrations of smoke or gases included in the air. Accordingly, in this section, calculation of pressure and flowrate is made only on air.

When a pressure difference $\Delta P$ across an opening or a flowpath is given, the mass flowrate $Q$ can be obtained generally from the following equation

$$Q = \alpha A \sqrt{2g\gamma|\Delta p|} \tag{1}$$

where

$\alpha$ = flow coefficient, $\alpha = 1/\sqrt{\xi}$,
$\xi = 1 + \zeta + \lambda(L/D)$,
$\zeta$ = pressure loss coefficient of flowpath (or opening),
$\lambda$ = frictional resistance coefficient of flowpath,
$L$ = length of flowpath,
$D$ = equivalent diameter of flowpath,
$A$ = sectional area of flowpath (or opening),
$g$ = gravitational acceleration, and
$\gamma$ = density of air.

Since there may be cases, such as those shown in Fig. 2, where the pressure difference at the opening in question is not uniform because of the temperature difference between the two spaces $i$ and $j$ on each side of the opening, Eq 1 is more generalized and is expressed in the following equation

$$Q = K\alpha B(h_2 - h_1)\sqrt{2g\gamma|\Delta P_a|} \tag{2}$$

where for $\gamma_i = \gamma_j$

$$K = 1, h_2 - h_1 = H_h - H_l, \Delta P_a = \Delta P_{ij}$$

$$\gamma = \gamma_i(\Delta P_{ij} > 0), \text{ or } \gamma = \gamma_j(\Delta P_{ij} < 0)$$

and where for $\gamma_i \neq \gamma_j$

$$K = \frac{2\sqrt{2}}{3} \sqrt{1 + \frac{n}{(1 + n)(1 + \sqrt{n})^2}}, \quad \text{where } n = h_1/h_2$$

$$\Delta P_a = \frac{h_1 + h_2}{2} \Delta\gamma_{ji}, \quad \text{where } \Delta\gamma_{ji} = \gamma_j - \gamma_i$$

The values of $h_1$, $h_2$, and $\gamma$ for $\gamma_i \neq \gamma_j$ in Eq 2 can be obtained from Table 3 and Fig. 2, depending on the height of the neutral plane $Y_N$ and on

TABLE 3—*Relationship between* $h_1$, $h_2$, $\gamma$ *and position of neutral plane.*

| Position of Neutral Plane | Flow Part of an Opening | $h_1$ | $h_2$ | $\gamma$ $\gamma_o > \gamma_i$ | $\gamma_o < \gamma_i$ |
|---|---|---|---|---|---|
| $Y_N \geqslant H_h$ | whole area | $Y_N - H_h$ | $Y_N - H_l$ | $\gamma_o$ | $\gamma_i$ |
| $H_l < Y_N < H_h$ | upper part | 0 | $H_h - Y_N$ | $\gamma_i$ | $\gamma_o$ |
|  | lower part | 0 | $Y_N - H_l$ | $\gamma_o$ | $\gamma_i$ |
| $H_l \leqslant Y_N$ | whole area | $H_l - Y_N$ | $H_h - Y_N$ | $\gamma_i$ | $\gamma_o$ |

whether the value of $\Delta\gamma_{ji}$ is positive or negative. In addition, according to Fig. 2, $Y_N$ is given by the following equation

$$Y_N = \Delta P_{ji}/\Delta\gamma_{ji}$$

When Room $i$ is considered as a subject, let $Q_{ji}$ be the mass rate of air

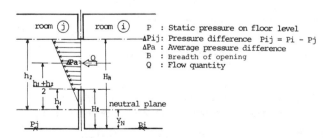

FIG. 2—*Mathematical model of flow quantity, pressure, and opening conditions.*

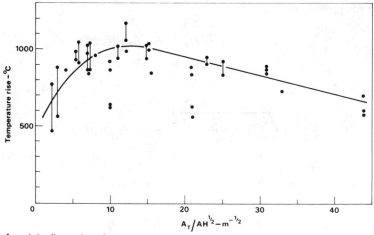

A$_T$ excludes floor and opening
Fire load densities in range 20-40kg/m² (for details see reference (3))
Various compartment shapes; fuel is 2cm x 2cm wood sticks, 2cm apart in cribs

FIG. 1—*Time mean temperature near ceiling.*

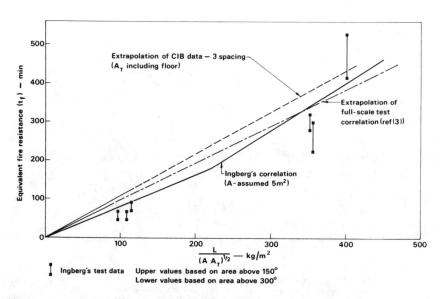

I  Ingberg's test data    Upper values based on area above 150°
Lower values based on area above 300°

FIG. 2—*Ingberg's recommendations and extrapolation of other data.*

First is the empirical relationship between $R$, the rate of burning, and $A\sqrt{H}$, the ventilation factor for window of area $A$ and height $H$

$$R = kA\sqrt{H} \qquad (1)$$

where $k$ is 5.5 kg m$^{-5/2}$ min$^{-1}$ or 0.09 kg$^{-1}$ m$^{-5/2}$ s$^{-1}$ approximately, and second is the heat balance

$$\dot{Q} = I_c + I_w + I_R \tag{2}$$

where

$\dot{Q}$ = rate of heat release,
$I_w$ = rate of heat loss to walls, etc.,
$I_c$ = rate of convected heat loss, and
$I_R$ = radiation (window) loss.

$I_R$ is generally negligible in its effect on the heat balance but is important for external hazard, and $I_w$ can accommodate the effects of the different thermal properties in the walls, a simple statement hiding the complexities of the analyses of Kawagoe and Sekine [7,8] and Magnusson and Thelandersson [9], which are the current bases for the design codes most widely recognized.

$\dot{Q}$ presents our first problem. $\dot{Q}$ is conventionally related to $R$ by an effective calorific value. Kawagoe and Sekine [7,8], by fitting calculations to data, found 2575 cal/g to be the best value for wood fires. By virtue of Eq 1, one can write $\dot{Q} \propto A\sqrt{H}$, and this, written directly, has in fact a more plausible basis than writing $\dot{Q} \propto R \propto A\sqrt{H}$ because, for ventilation-controlled fires, $A\sqrt{H}$ can be seen to physically determine the maximum possible heat release, while the burning rate is only empirically related to it. Magnusson and Thelandersson pursue the process of fitting calculations to data by developing shape factors for the variation of $\dot{Q}$ in time and normalizing the total heat release with that available from the calorific value of the fuel. Harmathy [6] has criticized the neglect of heat released outside but, so long as the effective calorific value is empirical, the significance of this neglect is not immediately apparent (see section on the fire outside). Calculations based on Eqs 1 and 2 for ventilation-controlled fires generally require greater protection for structures than fuel-controlled fires, and we have to ask whether there is any benefit from further refinements of theory. In what follows, a number of theoretical points are raised which are possibly not of great significance in relation to existing design procedures for calculating fire resistance requirements, but it must be remembered that providing adequate fire resistance is not our only problem.

1. A description of fire behavior adequate to describe an internal fire is insufficient for the designer of protection against fire spread up the outside of the buildings. The reason is that Eq 1 defines a rate of fuel loss which, even if it were a completely correct prediction of the rate of burning $R$, which strictly it is not [3,10], tells us nothing about how much unreacted fuel is emerging out of the window and contributing to the outside flames.

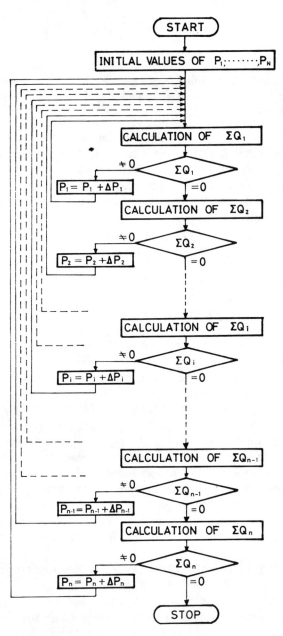

FIG. 3—*General procedure for calculation mass balance equations in a building.*

where

$V$ = air volume,

$\gamma$ = specific density of air, and

$\Delta C_i$ = a change of concentration in the time interval $\Delta t$ in the compartment, that is, $\Delta C_i = C_i(k + 1) - C_i(k)$

From the mass balance in the compartment, expressing $\Sigma Q_i \equiv \Sigma Q_{ji} = \Sigma Q_{ij}$, and $Q_{ij}$, $Q_{ji}$, etc., at $t = k\Delta t$, respectively, by $Q_{ij}(k)$, $Q_{ji}(k)$, etc., the following equation is obtained from Eq 8

$$C_i(k + 1) = \frac{\Sigma [Q_{ji}(k) \times C_j(k)]_j}{\Sigma Q_i(k)} + \left\{ C_i(k) - \frac{\Sigma [Q_{ji}(k) C_j(k)]_j}{\Sigma Q_i(k)} \right\} \times$$

$$\exp[- \frac{\Sigma Q_i(k)}{V_i \gamma_i(k)} \Delta t]$$

(9)

The value $C_i(k)$ of the smoke or a gas concentration in a compartment can be expressed in terms relative to the concentration of $C_F$ in a fire compartment, providing that $C_F$ can be assumed to be constant (for example $C_F = 1$) during the fire.

## Heat Calculation

Regarding heat calculation to obtain the temperature in each compartment, heat balance during a time interval $\Delta t$ at $t = k\Delta t$ in Compartment $i$ is considered in a manner similar to that described previously for the calculation of concentration. The heat balance equation is described by the equation

$$\left\{ C_p[\sum_j (Q_{ji} \times T_j) - \sum_j (Q_{ij} \times T_i)] - \sum_w [A_{iw} \times h_{iw}(T_i - T_{iw})] \right\} \Delta t = (V\gamma C_P \Delta T)_i$$

(10)

where

$C_P$ = specific heat of air at constant pressure,

$T_i$, $T_j$ = temperature in Compartments $i$ and $j$, respectively, where $j$ is a compartment connected with $i$,

$\Delta T$ = a change of temperature in a time interval $\Delta t$,

$A_{iw}$ = surface area of arbitrary wall (or ceiling or floor) $w$ in Compartment $i$,

$T_{iw}$ = surface temperature of the wall, and

$h_{iw}$ = heat transfer coefficient from Air $i$ to the Wall $w$.

For ventilation-controlled fires we have

$$\dot{Q} = \dot{q}\Delta H_{ox} \tag{8}$$

or an identically equal expression in terms of fuel, making use of the mass balances in Eq 4 and the values of $r$ and the fuel calorific value $\Delta H_f$. $\Delta h - c(T^* - T_0)$, the net enthalpy transfer to the fuel is denoted by $L$.

We obtain finally

$$\lambda_{ox,2} = \frac{Mn'}{M + R} - \frac{[\bar{h}A_T'/c + (M + R)]\theta}{(M + R)B} \tag{9}$$

$$\lambda_{f,2} = \frac{Rn''}{M + R} - \frac{[\bar{h}A_T'/c + (M + R)]\theta}{r(M + R)B} \tag{10}$$

where

$$\theta = \frac{E}{RT_0^2} \times (T_2 - T_0) \quad \text{and} \quad B = \frac{E}{cRT_0^2} \times \Delta H_{ox} \tag{11a}$$

$$n' = \lambda_{ox,1}\left(1 - \frac{RL}{\lambda_{ox,1}M\Delta H_{ox}}\right) \quad \text{and} \quad n'' = \left(1 - \frac{L}{r\Delta H_{ox}}\right) \tag{11b}$$

$$(Mn' - W\theta)\left(Rn'' - \frac{W\theta}{r}\right)\frac{ap^2e^{-E/RT}}{(M + R)^2} = W\theta + lR/B \tag{12}$$

and

$$W = \frac{\bar{h}A'_T/c + (M + R)}{B} \quad \text{and} \quad l = \frac{EL}{cRT_0^2}$$

The feedback $l$, while important in determining $R$, is not a major feature of the gaseous heat balance, so we shall discuss only

$$y = ap_0^2e^{-E/RT_0} = \frac{(M + R)^2 W\theta e^{-\theta/1 + B\theta}(1 + \theta/T_0)^2}{(Mn' - W\theta)(Rn'' - W\theta/r)} \tag{13}$$

where we write

$$e^{-E/RT_2} = e^{-E/RT_0}e^{-\theta/(1 + B\theta)} \tag{14}$$

where

$$B = RT_0/E$$

Equation 13 can produce the classic S-form well known, for example, in stirred reactor and self-heating systems limited by diffusion (Fig. 4), and recently referred to in connection with fire suppression by Williams [12]. For typical values of the various parameters, two stable and a third unstable equilibria exist (Curve $A$ which is drawn on the assumption that $rR > M\lambda_{ox,1}$ but is otherwise diagrammatic). The limits of stability are not necessarily exactly at the critical points. In general, the right hand side of Eq 13 is proportional to a power of scale between 2 and 3, while the left side is proportional to volume, that is, (scale)$^3$. The lower critical point exists only near $\theta$ of order unity. Thus, if we write the numerator as $\theta^m e^{-\theta}$ and $Mn'/W \gg m$ and $rR/W \gg m$, the lower critical value of $\theta$ is $m$, and, in general, we are concerned with temperatures much higher than are given by this. A discussion of fully developed fires can proceed without concern for this lower branch except for the one aspect we discuss next. Curves without critical points can occur, but they imply a low value of $Mn'/W$ or $R/W$, for example, a high wall loss or very small openings (Curve $B$). As drawn, the curve is a function of $Mn'/W$ and $rRn''/WB$ and $T_0$, and $R$ is assumed constant. Before discussing the further implications of this, we shall reexamine the conventional heat balance.

### Conventional Heat Balance

The conventional heat balance of a ventilation-controlled fire (the kind of fire in which the temperature is usually regarded as uniform), based on previous equations and $\lambda_{ox,2} = 0$, becomes formally

$$\theta_{max} = Mn'/W = \frac{Mn'B}{\bar{h}A_T'/c + (M + R)} \tag{15}$$

the upper limit shown in Fig. 4.

Introducing the time dependence of $\bar{h}$ by introducing the thermal coupling of the gases to the walls etc., leads in principle to the equations used by Kawagoe and Sekine [7,8] and Magnusson and Thelandersson [9].

My colleague, M. Bullen [13], has recently looked at the conventional heat balance equation with the addition of a conventional fuel surface energy balance for a pure liquid to determine $R$, and, with a more detailed treatment of emissivity for the radiation from the walls and from the flame, to define $\bar{h}$. With $M \propto A\sqrt{H}$, the heat generation has been taken as either $M\Delta H_{ox}$ or as $R\Delta H_f$, only the former being valid if $R > M\lambda_{ox,1}/r$

have been made on the control methods for smoke. However, because smoke movement is greatly influenced by external conditions (air temperature, wind), conditions of a building (spatial construction, opening conditions, temperature distribution in each space, etc.), it is difficult to ensure uniformly the safety of all individual buildings by means of a certain method of system. Therefore, by comparing the methods or systems posed for a building with a quantitative evaluation of their efficiencies under different conditions, the most suitable one may be chosen from them and the requirements for the method or system derived, for example, as to sizes of ventilation openings, amount of forced ventilation, and so on. For these reasons, some calculation methods are necessary for designing the smoke control systems.

Mainly to explain the calculation procedure use of the simplified method in more detail, a case study is introduced here, taking the following building model as an example.

## Model for Calculation

A mathematical model for the simplified method is shown in Fig. 4, focussing on the floor of origin, on which fire room ($F$), corridor ($C$),

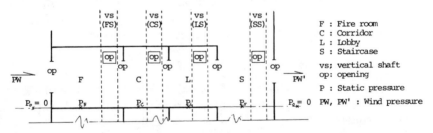

FIG. 4—*A model for simplified calculation* (*sectional elevation of fire floor*).

lobby ($L$), and staircase ($S$) are arranged in this order. Each of the compartments has two openings, one facing the open air and the other connecting the compartment to a vertical shaft (*FS, CS, LS, SS*). The vertical shaft here represents an air-conditioning duct system, elevator shaft, smoke tower, etc.

## Formation of Smoke Control

It is idealistic to confine the spread of smoke only to the compartment of origin and ensure steady safety of the occupants in the rest of the building from the smoke and gases in the event of fire.

The location of measures to cut off the smoke spread should be decided in relation to such factors as emergency evacuation systems. In this model, the openings or doors lying between *F-C, C-L,* and *L-S* can be considered as the prearranged position for stopping smoke on the fire floor, and all of these positions are taken into consideration in the calculation. On the other hand, as the prearranged position for forced ventilation, the compartments *F, C, L, S,* and the combinations of them can be considered as shown in Table 4, even if most of them are not practical. In the design of smoke control, it may be sufficient to choose some practicable combinations from among these four.

TABLE 4—*Combination of forced ventilations.*

| Number of Position for Forced Ventilation | Position for Forced Ventilation | | | |
|---|---|---|---|---|
| | Fire Room | Corridor | Lobby | Staircase |
| 1 | − | | | |
| | | + − | | |
| | | | + − | |
| | | | | + |
| 2 | − | + − | | |
| | − | | + − | |
| | − | | | + |
| | | + − | + − | |
| | | + − | | + |
| | | | + − | + |
| 3 | − | + − | + − | |
| | − | + − | | + |
| | − | | + − | + |
| | | + − | + − | + |
| 4 | − | + − | + − | + |

NOTE— + means air supply, − means smoke exhaust; +  − means + or −.

## Procedure of Calculation

While there are several simplified-method smoke control designs, a method which enables one to obtain very easily the required amount of forced ventilation will be introduced here as an example. In this example, only such principal spaces as the compartments (*F, C, L, S*) and the vertical shafts (*FS, CS, LS, SS*) are used for calculation. Vertical shafts and spaces generally have a significant effect on controlling the smoke movement. It is assumed in this model that each of the shafts is connected with a synthesized resistance of flow to the external air. The pressures on a standard level in these principal spaces are represented by $P_F$, $P_C$, $P_L$, $P_S$, $P_{FS}$, $P_{CS}$, $P_{LS}$, and $P_{SS}$. Table 5 shows the relation between each mass balance equation and the pressures concerned with the equation,

$M$ is not as sensitive to variation in high compartment temperatures as $R$ is expected to be; their characteristic variations with temperature are sketched in Fig. 6. At low mean temperatures, fuel volatiles may be evolved only by virtue of the local temperatures being high enough. Such a relationship can be incorporated into Eq 13.

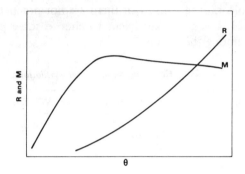

FIG. 6—*Typical dependence of* R *and* M *on temperature.*

Let

$$R = F\theta^s \tag{17}$$

where $F$ and $s$ are functions of the fuel type. Then

$$ae^{-E/RT_0} \propto \frac{\theta^{m-1} e^{-\theta/(1 + B\theta)}}{\left(\dfrac{Mn'}{W} - \theta\right)\left(n''F\theta^{s-1} - \dfrac{W}{R}\right)} \tag{18}$$

If $s > 1$, as suggested in Fig. 6, there is a minimum $\theta$ to which the preceding arguments apply unless local nonconformities are incorporated. The effect of such a dependence of $R$ on $\theta$ is to suppress the low temperature ignition of Fig. 4 (at $\theta$ of order unity) which corresponds to a few tens of degrees centigrade temperature rise. The minimum temperature rise must normally exceed this to produce finite fuel concentrations from solids and produce temperature characteristics of the kind sketched in Fig. 7. Although a full discussion of stability is outside the scope of this paper, the lower branch is readily seen to be unstable. If the system were to exist as a point on this branch, a fall in temperature would suppress fuel generation and heat production in the gaseous phase and so bring about extinction; a small perturbation upward would raise $\lambda_f$ from a low value and would raise the temperature, but would alter the oxygen concentration very little. More heat would be produced, and the system

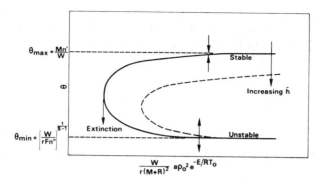

FIG. 7—*Uniform fire—role of kinetics* (R *temperature dependent*).

would necessarily move upwards to the upper branch. At the upper branch, the system is oxygen controlled, and a perturbation, which raises the oxygen concentration, lowers the temperature or vice versa. The lower branch is the *bb* portion of Bullen's solution of Fig. 5.

What is important for us is that, in this model, there is no way of reaching the upper stable state except by providing, in a transient stage, some energy for the fire to cross the lower boundary and "jump" to the upper one, a process which presumably corresponds in at least some aspects to "flashover". Nonuniformities allow $R$ to be finite even at temperatures below the mean corresponding to the lower branch.

Raising $F$ (or the rate of production of fuel at any given temperature) promotes the onset of instability and "flashover" in a developing fire (lowering the temperature of the lower asymptote), but hardly raises the temperature of the fully developed fires. This is because the excess fuel is mostly not burnt and acts as a heat sink.

Thus, increasing the mass transfer potentiality of the fuel by, for example, reducing its latent heat reduces the threshold to be reached by the growing fire and, one presumes, the rate of growth towards that threshold; that is, more flammable fuels produce faster growing fires and a more easily reached flashover.

## Low Values of $A\sqrt{H}$

It is clear that, because the basic temperature characteristics of $R$ and $M$ are different in form (Fig. 6), lowering the size of the opening could, in certain circumstances, lead to temperatures too low to sustain the fuel flow, and a ventilation-controlled regime at low temperature is in theory not possible. It can be shown that for $s > 1$, *aa* and *bb* in Fig. 5 curves derived from Eqs 15 and 17 intersect twice, if at all, producing a second pair of logically unacceptable solutions in which the basic assumptions of our model are inappropriate. For $s = 2$, the equations can readily be solved to give the condition for the existence of a ventilation-controlled regime ($R > M\lambda_{ox}/r$) and thereby two intersections as

$$F > \frac{4\bar{h}A_T'/c(1 + r/\lambda_{ox,1})}{(rB - 1)^2}$$

This condition is, in effect, a requirement imposing a limit on wall heat loss or a minimum surface area of fuel to which $F$ is proportional.

From Eqs 15 and 17, the limiting value of the air flow $M$ for ventilation-controlled fires, having low thermal feedback or high heats of reaction is

$$Mc \doteq \frac{\bar{h}A_T'(T_2 - T_0)}{\lambda_{ox,1}\Delta H_{ox}}$$

where $T_2$ is the temperature of the gases. Taking $\bar{h} = 0.01$ kW/m$^2$ [3] as a typical value, $T_2 - T_0 \approx 600\,°C$. $\lambda_{ox}\Delta H_{ox}$ as 4000 kJ/kg, we obtain

$$\frac{A\sqrt{H}}{A_T} = \frac{\bar{h}(T_2 - T_0)}{0.5\lambda_{ox}\Delta H_{ox}} \approx 0.003\,\text{m}^{1/2}$$

a result of the same order as Tewarson's lower limit for the normal ventilation-controlled regime, at which periodic minor explosions occur [11]. Physically, this implies that the heat released is virtually all lost to the walls. Attempts to reduce $M$ below the value at $Y$ must result in either extinction or some behavior which cannot be described by our assumption, for example, nonuniformity or transience (oscillations). In terms of Eq 18, this is because there is no solution for positive fuel and oxygen concentration of $\alpha < (W/rn''F)^{1/s-1}$.

### An Oscillatory Behavior

Let us return to considering Fig. 7 and a fire on its upper branch. If temperatures are high enough, the evolution of unburnt fuel is more significant than the evolution of unreacted oxygen, and the system is ventilation controlled.

Consider now a reduction in the ventilation factor $A\sqrt{H}$. $M$ falls, and the system equilibrium moves to a lower curve, eventually reaching one from which future reduction causes "extinction". In so far as there is some residual local burning, for example, below the surface of cracked fuel elements, there will be production of fuel gases at what, in theory, is an average temperature rise of zero; that is, the lower branch of the S-curve in Fig. 4 is not wholly suppressed as in Fig. 7, and the possibility of a reignition (for example, by self-heating) exists (as illustrated in Fig. 8). However, the curve involved is changed because the value of $\bar{h}$ changes; in effect, it becomes negative as heat from the hot walls returns to the gases.

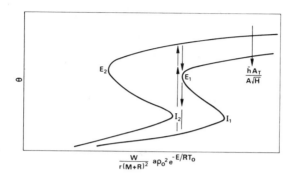

FIG. 8—*Possible oscillation caused by decreasing* $A\sqrt{H}$.

If the drop in $\bar{h}$ is large enough for $E_1$ to be the right of $I_2$, the system would reignite and possibly oscillate (see Fig. 8). A transient analysis would be needed to explore such behavior. This, of course, is not the only possible source of oscillation.

Tewarson [11] has drawn attention to the possible complex roles of kinetics at low ventilation; the fact that Tewarson quotes the same critical for two different fuels supports the view that this term is not primarily related to the fuel.

This argument is presented here only in the sense of such a suggestion, but it tends to confirm the observed insensitivity of the behavior to a change of fuel and suggests that we need not be concerned with the critical window openings becoming more in the range of typical building values and so presenting new hazards arising from instability as such just because the fuel has changed.

## The Fire Outside

### The External Flame

One of the remaining problems of building design is to provide protection from flames emerging from openings in building facades. Calculating the heat transfer, the temperatures and velocities of the hot gases are part of the problem of estimating the requirements for external structures and of reducing spread externally up the facade. Of course, the radiation depends on the emissivity, and this cannot yet be calculated *a priori* for buoyant, turbulent, diffusion flames. The safest and most practical procedure is to assume a radiation flux level.

The temperature of the gases leaving the window—or perhaps, more exactly, approaching the window—is $T_2$ but, as soon as the plume turns upwards, entrainment becomes faster, and combustion and dilution processes commence. The temperature at which flames radiate is clearly determined by these processes as much as—and probably more than—by the temperature $T_2$. Because of this, we have to admit that there is no basic theory for the external flame which can be expressed in terms of building and occupancy characteristics, such as fire load and window opening.

The external flame might be regarded roughly as a region in which the stoichiometric air requirement is finally met. The differences in temperature between a wholly open free-burning flame and one in the flame external to a compartment will depend upon the differences in the heat loss, the loss from the former being largely radiation to cold surroundings, and from the other radiation to walls and fuel etc; so long, therefore, as these are not too hot, the difference will be small. Stromdahl [16] reports radiating temperatures of flames as about 1000°C some 200°C higher than in the room. We also need comparisons with "open" fires since some external flames may be more easily considered as flames from such fires perturbed by compartments than as extensions of compartment fires.

### The External Heat Release

Harmathy [6] has suggested evaluating the heat release within the compartment by exploiting published flame correlations to estimate the external heat release. For this to be superior to basing heat release on air flow (even if one has to introduce some allowance for the exit oxygen), one has (a) to neglect certain difficulties [17] in the usage of the correlations, and (b) assume that $R/A\sqrt{H}$ does not vary from 0.09 (5.5 kg min$^{-1}$ m$^{-5/2}$) so that one can calculate flame length in terms of $A\sqrt{H}$.

The total heat release rate is $R\Delta H_f$ where $\Delta H_f$ depends on the degree of conversion to carbon dioxide and water. Within the limits set by the simplified kinetics in this model, the rate of heat release outside, as a fraction of the whole, is

$$f = 1 - \left\{ \frac{M\lambda_{ox,1} - (M + R)\lambda_{ox,2}}{rR} \right\}$$

which from Eq 9

$$f = 1 - \frac{c\theta}{\Delta H_{ox} r} \left( \frac{M + R}{R} \right) \frac{RT_0^2}{E} - \frac{\bar{h}A_{T'}}{\Delta H_{ox} r} \left( \frac{\theta}{R} \right) \frac{RT_0^2}{E} - \frac{L}{\Delta H_{ox} r} \quad (19)$$

that is, from Eq 11$a$

$$f = 1 - \frac{\theta}{rB}\left(\frac{M + R}{R}\right) - \frac{\bar{h}A_T'}{B}\frac{\theta}{R} - \frac{L}{\Delta H_{ox}r}$$

For values of

$$r\Delta H_{ox} = \Delta H_f = 18 \text{ MJ/kg}$$

$$\theta \approx 1000\,°\text{C}$$

$$M/R \approx 5$$

$$\bar{h} = 10^{-2}\,\text{kW/m}^2\,°\text{C}$$

$$A\sqrt{H}/A_T \approx 0.05 \text{ m}^{-½}$$

we have

$$f = 0.40 - \frac{L}{\Delta H_f}$$

very close (perhaps fortuitously) to Harmathy's estimate (based on a lower $\Delta H_f$ and zero $\Delta h$), though clearly the agreement of numerical values, being based on correlations experimental but different, but neither wholly theoretical, is to be expected and, for typical variations about these values, values of from, say, 0.2 to 0.6 could be envisaged.

Without a determination of $R$ independently of $M$, one cannot proceed far in a discussion of Eq 19. We note that reducing $L$ has two effects on $f$, one directly on the last term and the other in increasing $R$, and, if this is of greater importance, $f$ is increased.

## Conclusion

A number of diverse aspects of compartment fires have been examined briefly, and attention is drawn to certain secondary features of these fires which lie outside what is acceptable as a theory for the design prediction of fire resistance. One can describe one feature of flashover—perhaps only of certain kinds of flashover—as a transition across a fuel-controlled regime which is unstable in conditions of uniform burning. The descriptions employed here are highly oversimplified and the conclusions speculative, but future studies along these lines might be rewarding. Some theoretical basis has been indicated for the regimes of very low $A\sqrt{H}/A_T$ departing from that of being conventionally ventilation controlled. A calculation of a critical parameter gives a value of similar order to that observed by Tewarson and suggests an important role for wall heat loss. Certainly, once one departs from the design of fire resistance, the theory of fully developed fires requires considerable development to provide an adequate base for the other fire problems, be they in design, fire stability and control, or the role of extended flammable surfaces in producing fuel which burns outside the window.

*Acknowledgment*

This paper is Crown copyright. It is reproduced by permission of the Controller, Her Britannic Majesty's Stationery Office. It is contributed by permission of the Director of the Building Research Establishment. The Fire Research Station is the Joint Fire Research Organization of the Fire Offices' Committee and Department of the Environment.

**References**

[*1*] Ingberg, S. H., *National Fire Protection Association Quarterly*, Vol. 22, 1928.

[2] Robertson, A. F. and Gross, D. in *Fire Test Performance, ASTM STP 464*, American Society for Testing and Materials, 1970, pp. 3–29.

[*3*] Fujita, K., "Characteristics of Fire Inside a Non-combustible Room and Prevention of Fire Damage," Report No. 2(h), Japanese Building Research Institute, Tokyo.

[*4*] Kawagoe, K., "Fire Behaviour in Rooms," Report No. 27, Japanese Building Research Institute, Tokyo, 1958.

[*5*] Thomas, P. H. and Heselden, A. J. M., "Fully Developed Fires in Single Compartments. A Co-operative Research Programme of the Counseil International du Bâtiment," Fire Research Note 923/1972, Joint Fire Research Organization.

[*6*] Harmathy, T. Z., *Fire Technology*, Vol. 8, No. 3, 1972, pp. 196–217; *Fire Technology*, Vol. 8, No. 4, 1972, pp. 326–351.

[*7*] Sekine, T., "Room Temperature in Fire of a Fire Resistive Room," Report No. 29, Japanese Building Research Institute, Tokyo, 1959.

[*8*] Kawagoe, K. and Sekine, T., "Estimation of Fire Temperature-Time Curve in Rooms," Japanese Building Research Institute Occasional Reports No. 11, Tokyo, 1963; and Report No. 17, Tokyo, 1964.

[*9*] Magnusson, S. and Thelandersson, S., "Comments on Rate of Gas Flow and Rate of Burning for Fires in Enclosures," Bulletin 19, Division of Structural Mechanics and Concrete Construction, Lund Institute of Technology, 1971.

[*10*] Gross, D. and Robertson, A. F., "Experimental Fires in Enclosures," Tenth Symposium (International) on Combustion, The Combustion Institute, 1965, pp. 931–942.

[*11*] Tewarson, A., *Combustion and Flame*, Vol. 19, No. 3, 1972, pp. 101–111; and Vol. 19, No. 3, 1972, pp. 363–371.

[*12*] Williams, F. A., *Journal of Fire and Flammability*, Vol. 5, 1974, pp. 54–63.

[*13*] Bullen, M., Personal communication.

[*14*] Thomas, P. H. and Nilsson, L., "Fully Developed Compartment Fires: New Correlations of Burning Rates," Fire Research Note 979, Fire Research Station, Borehamwood, England, 1973.

[*15*] Nilsson, L., "The Effect of Porosity and Air Flow Factor on the Rate of Burning for Fire in Enclosed Space," Report No. R22, Swedish National Building Research Institute, 1971.

[*16*] Stromdahl, I., "The Tranas Fire Tests: Field Studies of Heat Radiation from Fires in Timber Structures," Document D3, Statens Institute för Byggnadsforskning, Stockholm, 1972.

[17] Thomas, P. H., *Fire Technology,* Vol. 11, No. 1, 1975, pp. 42–48; and *Fire Technology,* Vol. 11, No. 2, pp. 143–145.

# DISCUSSION

*R. A. Lacosse*[1] (*written discussion*)—Dr. Thomas showed a slide of a fire using insulating board. Reference was made of using *unfinished* insulating board.

So as not to raise questions and confuse those not knowledgeable in this area, this is to advise that, since the late 1950's, all insulating board products used as exposed interior finishes have been manufactured with a flame resistant finish.

For this application, there are presently *no* unfinished insulating boards in this country (United States) and should be so recorded in Dr. Thomas' paper.

*P. H. Thomas* (*author's written closure*)—Mr. Lacosse refers to "flame resistant" finishes. Of course, if the lining really did "resist" flame, there would, in effect, be no flammable lining, but if the finish, as is more usual, only retards flame and thereby prevents some fires and perhaps delays flashover in others, it may not necessarily burn any less fast in a fully developed fire should this occur. In so far as it does burn less rapidly than the untreated material used in the experiments showing large flames out of the window, because of its treatment or greater conductance or any other reason, the flames will presumably be smaller; the difference would be one of degree unless flashover is actually prevented.

*R. G. Gann*[2] (*written discussion*)—I infer from your treatment of fully developed fires in ventilated compartments that the behavior of these fires is essentially independent of the preflashover conditions. That is, the combustion depends primarily on the fuel character and the ventilation rate. To what extent is this realistic? It seems that, as we tend towards less flammable materials (that is, as we attain more success on that front), a greater degree of sophistication must be applied in modeling the transition from local to widespread burning.

*P. H. Thomas* (*author's written closure*)—Dr. Gann is correct in inferring that I am assuming here that the behavior after flashover is independent of the behavior before flashover. This would not be so, for example, if fire wholly burns out part of fuel before flashover has occurred. The flashover period may be influenced by many factors having little or no effect on the fully developed fire, and it is doubtful if design for resist-

[1] Manager, Technical and Manufacturing Services, Acoustical and Board Products Association, Parkridge, Ill. 60068.
[2] National Bureau of Standards, Washington, D.C. 20234.

ing fully developed fires should be based on anything but the worst case, so long as this has a high probability.

There are situations where a change of occupancy can turn a situation in which flashover is virtually impossible into one in which it is very probable, and this possibility is recognized in principle, even if implicitly, by some regulatory bodies in the issue of building licenses and the variation of requirements for different occupancies.

*James Quintiere*[1]

# Growth of Fire in Building Compartments

**REFERENCE:** Quintiere, J., "**Growth of Fire in Building Compartments,**" *Fire Standards and Safety, ASTM STP 614,* A. F. Robertson, Ed., American Society for Testing and Materials, 1977, pp. 131–167.

**ABSTRACT:** A review was made of both full-scale and scale model experiments concerned with fire growth and spread in building compartments. It appears that "flashover," that is, the rapid transition to a fully developed room fire, could be initiated by a fully involved chair fire alone or by a large waste container ignition source against a combustible lining material. Scale model results continue to provide valuable insight, but the validity of partial scaling results must be considered for each type of experiment. A quasi-steady idealized mathematical model was developed to analyze the various parameters of fire development in a room. These theoretical results show the significance of fuel properties, fire size and location, room and doorway dimensions, and wall thermal properties. The limitations of a mathematical approach are also discussed.

**KEY WORDS:** fires, flashover, literature reviews, scale models, mathematical models, fluid flow, heat transfer, theories

## Nomenclature

$A$    Surface area

$c$    Orifice coefficient

$C, C_g$    Specific heat of solid, fluid

$d$    Density ratio, $\varrho_g/\varrho_a$

$D$    Horizontal distance from flame

Fr    Froude number

$F_{dF}$    Geometric configuration factor between the thermal discontinuity plane and the floor

$g$    Gravitational acceleration

Gr    Grashof number

---

[1] Mechanical engineer, National Bureau of Standards, Washington, D.C. 20234.

$h$    Convective or total heat transfer coefficient
$H$    Height
$\Delta H$    Heat of combustion
$\Delta H_v$    Heat of volatilization
$i$    Enthalpy
$k$    Thermal conductivity
$k_e$    Entrainment constant
$k_0, k_1, k_2, k_f$    Constants
$K$    Wall and ceiling conductance length scale
$L$    Length of room
$\dot{m}$    Rate of mass flow
$p$    Pressure
$\dot{q}$    Rate of heat flow
$r$    Mass air to fuel ratio
$R_f$    Radius of flame
$R$    Universal gas constant
$t$    Time
$T$    Temperature
$V$    Velocity
$W$    Width of room
$X_n$    Height of neutral plane
$X_d$    Height of thermal discontinuity
$\alpha$    Coefficient of volume expression
$\beta$    Parameter defined by Eq 7c
$\gamma$    Area ratio, $A_F/A_W$
$\delta$    Wall thickness
$\delta_f$    Effective heat transfer length ahead of flame
$\Delta$    Difference operator
$\varepsilon$    Emissivity
$\mu$    Viscosity
$\varrho$    Density
$\tau$    Dimensionless time, $th^2/k\varrho c$
$\omega$    Parameter defined by Eq 7b

*Subscripts*

$a$    Air
$b$    Burning
$d$    Thermal discontinuity
$e$    Entrainment
$f$    Flame
fuel    Fuel
$F$    Floor

    $g$    Hot gaseous combustion products
    ig    Ignition
    $n$    Neutral plane
    $o$    Doorway
    $p$    Plume
prod    Products
    py    Pyrolysis
    $r$    Radiation
    $s$    Fuel surface
    $v$    Volatilization
    $w$    Hot walls and ceiling
    $\infty$    Reference state

*Superscript*

( )″    Per unit area

The discovery of fire by mankind has been both a blessing and a curse. The term "combustion" connotes the beneficial aspects, such as heating our homes and yielding a useful energy source. On the other hand, the term "fire" implies a hazard and an uncontrollable state. In modern society, the amount of furnishings in residential occupancies has risen, increasing the chance of fire growth. Whether a small fire will become large is a question for the fire safety engineer. A definitive answer to this question cannot be given with our current state of knowledge. What can be given at this time is a review of what we know (or think we understand) about the spread of fire in enclosures.

This presentation will focus on the problem of fire growth within an enclosure or room. In this context, envision a room with openings, such as doorways or windows, within which an ignition has occurred. The ignition source or primary fire may grow in size due to heat transfer from the flames to its immediate surroundings. Air flows into the fire by the buoyant force of the fire itself, and the flow is limited by the size of the openings to the room. The fire may go out because its fuel is depleted or it may continue to grow. At some point in its growth, the room will be sufficiently heated to increase the rate of heat transfer to the fire and unburnt materials. Thus, the fire will tend to accelerate in growth due to this feedback process. The maximum fire size that can develop under this feedback process is limited by the air supply rate through the openings or the amount of available fuel in the room. This maximum fire size is usually termed the fully developed fire. The rapid fire growth process preceding this stage is termed "flashover." Several criteria may be used to define flashover, but a satisfactory definition is difficult to make in quantitative terms. Let it suffice that flashover is a critical state in the fire growth in a room which marks the difference between a fire that stays relatively small and confined to its initial surroundings and a fire that reaches its fully developed state.

## Review of the Literature

Recent reviews on the fire growth in building compartments have been given by Thomas [1][2] and Friedman [2]. Hopefully, the present discussion will add some additional information to a very multifaceted problem. However, the present review has been far from exhaustive, and further efforts to probe the literature and synthesize the facts will, I am sure, be beneficial.

### Full-Scale Fire Experiments

Efforts to understand the fire development in a room have invariably led to full-scale experiments of fires set by scientists in rooms of various shapes and contents. The motivation for these tests may have varied, but, in each case, the study has enabled the experimenter to observe the chronology of the fire development and to correlate this with various measurements. In retrospect, we can learn something about the way the fire has spread and the factors indicating the hazard.

It is known that accidental fires frequently start from a cigarette carelessly forgotton on an upholstered chair or bed. This type of ignition usually leads to a smoldering fire. The dynamics of the smoldering fire can be different from flaming combustion. The temperatures in the room will be relatively low, and the fluid flow of gases in the room may involve layering of the gases. It appears that carbon monoxide (CO) is the principal hazard of the smoldering fire. If the smoldering material is a polyurethane foam, a sufficiently strong airflow can induce flaming combustion; however, this appears to be an uncommon or rare occurrence [3]. The outbreak of flaming combustion in other materials, such as cotton padding in upholstered furnishings, is more likely. Hafer and Yuill [4] performed extensive experiments of flaming and cigarette ignitions on a variety of bedding and upholstered chairs. The developed fires were smoldering or limited in flaming combustion, probably by the very small window openings they had in the room. Had the windows been opened wider, it is expected that larger fires would have developed. In fact, it is remotely possible that flammable vapors which would collect in an enclosed room due to a smoldering fire could cause a very violent fire or explosion [5] upon the sudden addition of fresh air or a flaming ignition source.

---

[2] The italic numbers in brackets refer to the list of references appended to this paper.

In experimental studies, emphasis has been placed on residential occupancy fires such as the office, the living room, or the bedroom. In 1939, Ingberg [6] studied the fire severity and duration of typical living room furnishings. For a moderate amount of living room furnishings, the extent of fire spread was to some furniture items; whereas, for a heavier fuel loading, the living room fire led to flashover in the adjoining rooms. They concluded that the orientation of the furnishings is important with respect to the fire origin. Bruce [7] concluded, after a series of room fire experiments, that the direction of the flames on the first item ignited had an effect on the rate of fire development. Thus, the effect of fire location in the room and air currents which direct the flame of a small ignition source play a significant role in the early growth of a fire.

Waterman [8] conducted an extensive series of experiments to determine the conditions supporting room flashover. The room was 3.64 by 3.64 by 2.43 m high with inner surfaces of silica-asbestos board. He examined fires started on various items of furniture and also varied the size of the window openings to the room. He concluded that the mechanism for room flashover is not the ignition of the unburnt gases which collect along the ceiling but is due to combustible items being heated by radiation and convection to the point at which they will ignite. Moreover, in these experiments, he indicated that most of this radiation originated from the heated upper walls and ceiling and not directly from the flaming item. When any major furniture item became fully involved in flames, flashover generally followed. This usually occurred for burning rates $\geqslant 40$ g/s. For example, an upholstered chair fire could cause the fire to spread throughout the room by releasing a significant amount of energy to heat the room. The confined hot gases and heated walls and ceiling then radiate back to other items causing them eventually to ignite.

Developing fires in a bedroom ensemble have been reported by Hillenbrand and Wray [9], Croce and Emmons [10], and Croce [11]. The first study is noted for its demonstration of the effect of fire retardant materials on fire spread and for the unusual flow conditions that resulted at the doorway of the bedroom. The latter studies [10,11] are noted for the extensive analysis of the data by a team of scientists. These studies point out the importance of radiation from the smoke layer to the lower portions of the room. For example, Alpert in Ref 11 reports, just prior to flashover, a heat flux of 4 W/cm² to a horizontally mounted sensor (Gardon) located at the edge of the primary fire source (a burning mattress). He estimated that about 1 W/cm² was due to the hot smoke layer. Moreover, Orloff and de Ris in Ref 11 demonstrated that the ignition of a remote item (a bureau) was probably due to radiation received from the smoke layer.

Fires involving wood cribs and upholstered and foam chairs in rooms were studied by Hägglund, Jansson, and Onnermark [12]. They settled on a definition of flashover as the point when flames emerged from the doorway of their concrete test room (2.90 by 3.75 by 2.70 m high). This appeared to correlate with a temperature near the ceiling of 600 °C. Their data for wood cribs is represented in Fig. 1. It illustrates the classical

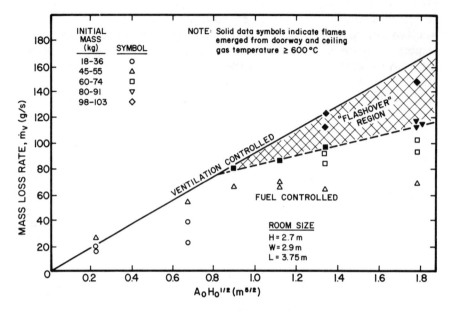

FIG. 1—*Mass loss rate of wood cribs in a room with various window openings* [12].

presentation of ventilation and fuel-controlled fully developed fires. It further illustrates a mapping of their flashover region on this plot. Thus, the results display the potential for flashover as a function of the window opening and amount of fuel involved in combustion. A generalized version of such a plot would certainly be welcomed by the fire safety engineer.

The importance of the initial or primary source fire on flashover potential has prompted some investigators to study individual items in a room enclosure. Gross and Fang [13] studied the burning characteristics of waste containers (18 to 120 litres) and upholstered chairs. They found the flames from these burning items would impart a heat flux of about 0.5 to 5 W/cm² to adjoining items. Fang [14] obtained similar heat flux results for a more extensive study of upholstered chair fires in rooms. The average duration of these chair fires was about 1 h. In most of these experiments, the maximum separation distance to cause ignition of adjacent paper, cotton cloth, or wood targets was 0.15 m. Theobald [15] also found that radiation from a burning chair could ignite cotton cloth 0.15 m away, but a burning wardrobe could ignite cotton cloth 1.2 m away.

Thus, the larger the room fire, the greater the radiant ignition hazard to the surroundings.

A summary of typical burning rates for various room items is shown in Fig. 2. Also shown in the figure are values for fully developed room

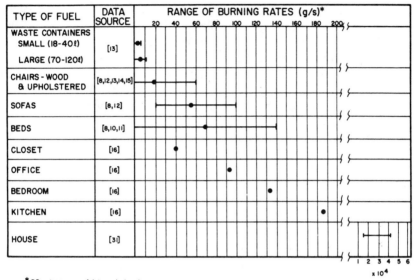

| TYPE OF FUEL | DATA SOURCE | RANGE OF BURNING RATES (g/s)* |
|---|---|---|
| WASTE CONTAINERS SMALL (18-40ℓ) LARGE (70-120ℓ) | [13] | |
| CHAIRS - WOOD & UPHOLSTERED | [8,12,13,14,15] | |
| SOFAS | [8,12] | |
| BEDS | [8,10,11] | |
| CLOSET | [16] | |
| OFFICE | [16] | |
| BEDROOM | [16] | |
| KITCHEN | [16] | |
| HOUSE | [31] | |

*●Denotes average of data or single value

FIG. 2—*Range of maximum burning rates for various residential combustible materials.*

fires estimated from Christian and Waterman [16]. In a room environment, the burning rate, or, more directly, the energy release rate of the burning item, will be significant with regard to the potential of flashover. Figure 2 gives a perspective on the maximum burning rate for typical furnishings. These data were collected from various sources, and, consequently, the enclosure size was not the same for all cases. However, the data were selected for fires which were not ventilation controlled.

Although a large burning item of furniture alone may be sufficient to cause room flashover, combustible lining materials along the walls and ceiling can often be ignited by relatively small ignition sources and thus contribute significantly to the development of flashover. Bruce [7] observed that the lining materials did not significantly get involved in fire until flashover occurred. His ignition sources were in the center of the room. However, Fang [17] studied lining materials exposed to a corner ignition source (a wood crib with a burning rate of approximately 7 g/s). He designated flashover by the radiative ignition of horizontal targets of newsprint and 6-mm-thick plywood located at the center of the room 10 cm above the floor. The newsprint ignited at a minimum of 1.7 W/cm², compared with 2.1 W/cm² which was required for the plywood. Fang related this to an average upper gas temperature of roughly 610°C. For his ignition source and wall linings alone (that is, a noncombustible ceiling and floor), he found that flashover occurred for wall materials tested which had an ASTM Test for Surface Burning Characteristics of Building Materials (E 84-75) flame spread classification (FSC) >180 or alternately an ASTM Test for Surface Flammability of Materials Using a Radiant Heat Energy Source (E 162-75) flame spread index (I) >150.

Similar experiments were conducted at Underwriters Laboratory [18] by Castino and co-workers. They investigated cellular plastics and other materials which lined the walls and ceiling of a room. When they subjected a room to a 9-kg (20-lb) wood crib corner fire (with an estimated burning rate of 6 to 10 g/s, which can be interpreted as a large waste container fire or a chair fire) 5 out of 6 lining materials tested became 90 to 100 percent involved in burning. These materials had ASTM Test E 84-75 FSC of 3 to 178. The paper surface on plaster board linings (FSC = 13) became up to 65 percent involved in flaming for a short time under the fire exposure of a 23-kg (50-lb) crib (with an estimated burning rate at 20 to 25 g/s). It appears that it does not take a very large corner fire to fully involve wall lining materials in a room.

In contrast to wall and ceiling linings, floors and floor covering materials seem to present a less severe problem. For example, Davis and Tu [19] report no more than 18 percent involvement of carpets subjected to corner fires of upholstered chairs or 6 to 30-kg wood cribs when the walls and ceiling of the room were noncombustible.

The role of lining materials and their contribution to room fire development is a complex phenomenon. The contribution of flames under an incombustible ceiling [20] and a combustible ceiling lining [21] have been studied by Hinkley, Wraight, and Theobald and further discussed by Thomas [22]. A qualitative sketch taken from these studies is displayed in Fig. 3. The contribution of direct flame radiation and combustible ceiling are shown in terms of the incident radiative heat flux to the floor and distance from the fire. Hinkley and Wraight [21] found that a combustible ceiling could increase the heat flux by more than twice that of an incombustible ceiling. Thomas [22] indicates that the time to heat up and ignite a combustible material is directly proportional to its thermal property $k\varrho c$ (the product of thermal conductivity, density, and specific heat). Hence, low density linings will tend to heat up quickly and ignite quickly.

Although the emphasis of this paper is the developing fire in a room, some other cases of building fire spread will be mentioned. Once a room fire has developed to a large size, its chance of spread to the surroundings increases. One hazard is fire spread into an adjoining corridor. The resultant floor and ceiling incident heat fluxes due to a room fire of various sizes are shown for a corridor in Fig. 4. These results were taken in the

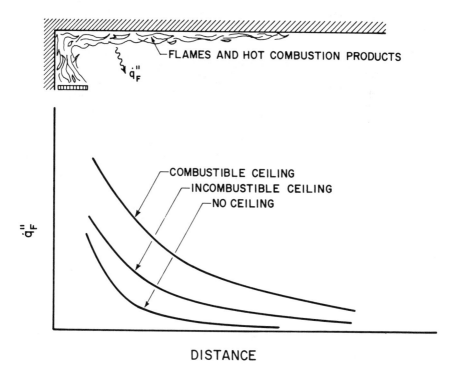

FLAMES AND HOT COMBUSTION PRODUCTS

$\dot{q}_F''$

COMBUSTIBLE CEILING

INCOMBUSTIBLE CEILING

NO CEILING

$\dot{q}_F''$

DISTANCE

FIG. 3—*Qualitative radiative heat flux distribution along a floor from flames and a heated ceiling [21,22,23].*

facility previously described by this author [23] in discussing the phenomenon of fire spread along corridor floor coverings. The results shown in Fig. 4 are for noncombustible corridor linings. The heat flux data are relative to cold surfaces, that is, to water-cooled heat flux gages. The ceiling flux is primarily convective, while the floor flux is radiative. It is significant that the results display the effect of external room vents (which supply additional air) and fire location for the four crib cases. Fire spread from a room to a corridor with combustible wall and ceiling linings was studied by several investigators [24–26]. The case of fire spread in combustible passages has been studied [27–29] with the application to mine fires in which the gas flow is totally in one direction. If enough fuel is available, these fires can become fuel rich, and the propagation then depends on the inlet air speed. This undirectional flow case can have relevance to vertical ducts and stairwells in building fires. The problem of flame projection from a window and possible spread to the next story of the building and the hazard of the fire plume from a burning house has been analyzed very extensively by Yokoi [30]. Finally, the dynamics of fire spread throughout wood-frame buildings has been studied by Butler,

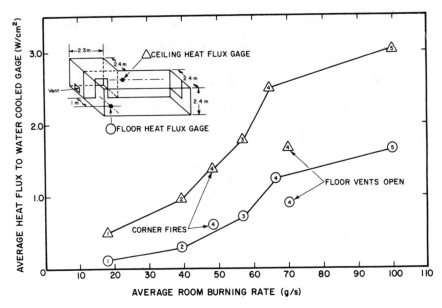

FIG. 4—*Incident heat flux to cold surfaces of a ceiling and floor in a corridor adjoining a room fire of various intensities. Numbers in symbols indicate the number of 18 kg wood crib fires in the room. Cribs were located in the center of the room and room vents were closed unless noted otherwise.*

Martin, and Wiersma [31], and the St. Lawrence burn experiments examined the potential of fire spread between buildings [32].

### Small-Scale Laboratory Studies

An international cooperative study examined the effect of many variables on flame spread through wood cribs within a 1-m-high compartment [33,34]. The initial conclusions from these studies show that the fire spread rate depends on the bulk density of the fuel during its early growth; then it accelerates as the compartment is heated. The spread rate was found to depend on the position and size of the ignition source, fuel height, stick spacing, and lining material. The spread rate had no dependence on compartment shape and little dependence on opening size. It is interesting that initial rates of spread differed significantly among the various laboratories. The data from these experiments will probably continue to be analyzed.

There is a strong desire in fire research to conduct scale-model simulations of full-scale experiments. The motivation for this is the obvious savings in cost and setup time and the easier ability to change variables. Although similitude modeling can have a mathematical basis, it is impossible to scale all of the dominant dimensionless groups in a fire experiment. For excellent dissertations on the mathematical basis of similitude modeling in combustion, one should refer to Spalding [*35*], Hottel [*36*], or Williams [*37*]. Both Hottel and Williams indicate the advantages of utilizing a high pressure chamber, and Williams goes further to indicate the added advantage of using a high pressure centrifuge. The "pressure modeling" approach primarily lies in adjusting the Grashof number (Gr) by increasing the ambient pressure such that the dimensionless mass transfer or burning rate is preserved since essentially

$$\frac{\dot{m}_b{}''l}{\mu} = f(\text{Gr}) \tag{1a}$$

$$\text{Gr} = \frac{gl^3(\Delta\varrho/\varrho)}{(\mu/\varrho)^2} = \frac{gl^3\beta\Delta T}{\mu^2}\left(\frac{p}{RT}\right)^2 \tag{1b}$$

for a perfect gas. Also, the convective heat transfer is preserved since the Nusselt number, $hl/k$, is also strongly dependent on Gr. The Froude number (Fr) is also preserved in the pressure modeling scheme, that is

$$\text{Fr} = \frac{\varrho V^2}{lg\Delta\varrho} \tag{2}$$

provided the imposed flow speed is adjusted. By preserving the Gr and Fr numbers in the pressure modeling method, the viscous, buoyancy, and convective processes will remain similar between the model and the prototype. The success of this method has been demonstrated by several investigators [*38–40*] for simple geometries and for wood crib fires in enclosures. The inability of this method to account properly for radiation emitted from heated surfaces and perhaps also for flame radiation is a pitfall of this approach.

A more commonly used scaling approach has been to burn "scaled-down" models at normal atmospheric pressure. It can be referred to as "Froude number modeling" since the Fr number is the primary variable preserved. However, in this modeling context it is convenient to write the Fr number as

$$\text{Fr} = \left[\frac{\dot{q}}{\varrho_\infty C_p T_\infty l^{5/2} g^{1/2}}\right]^2 \frac{(T/T_\infty)^2}{(T/T_\infty - 1)^3} \quad \text{(for a perfect gas)} \tag{3}$$

where the energy input rate $\dot{q}$ has essentially eliminated $V$ by a conservation of energy relationship given as

$$\dot{q} = \varrho C_p V l^2 \Delta T \tag{4}$$

Many investigators [41-47] have used this approach with varying subsidiary conditions which depended on the manner of representing the energy release rate $\dot{q}$ and the restriction of maintaining the ratio of fuel generated rate ($\dot{m}_v$) and airflow rate ($\dot{m}_a$) constant between the model and the prototype. The airflow rate is modeled as the product of door width ($W_o$) and height to the 3/2 power ($H_o^{3/2}$). Wall properties are usually modeled in this method in an attempt to maintain identical temperatures in scaling; however, radiation effects from flames and gases can upset this type of model. In general, the more complex the fire problem, the more likely that Fr number scale modeling will fall into an art form rather than one that can be supported by mathematics. Although scaling results may achieve a level of success, it may not be known whether nonscaled effects have compensated the results to give good agreement between the model and full scale. It is interesting to note that Hird and Fischl [48] used 1/10, 1/5, and 1/2 geometric scale models of a room and its furnishings with the full-scale furniture thickness maintained. Their scale model temperature and time to flashover results were in good agreement with the full-scale experiments.

### Mathematical Models and Correlations

It should be realized that, although experimental studies of room fire growth are valuable, they alone are not sufficient to provide us with a general understanding of the effect of room and fuel initial conditions. The layman probably could list quite well the number of variables which will affect the fire growth, and the experimenter will undoubtedly incorporate some of these in his fire tests. However, it is the mathematician that can vary all the governing parameters through the solution of equations which represent the fire growth. Although the mathematical approach has the most potential, its accuracy will only be as good as a correct understanding of the fire growth processes and the ability to model reality. Without a mathematical model, the fire safety engineer will have to rely on his impressions of full-scale fire experiments and relative material rankings from flammability test methods.

For the room fire problem, mathematical modeling has been motivated by the desire to generalize the results of experiments designed to predict the fire severity to the structure and the fire duration. Kawagoe [49] and Sekine [50] have set the tone of this approach by modeling the buoyancy driven airflow as an orifice flow problem through the room doorway or window. A global energy balance was applied to the room, and transient heat losses were accounted for in terms of the wall material properties. Thomas, Heselden, and Law [51] presented a comprehensive account of the fluid flow and heat transfer processes affecting fires in compartments with both large and small openings. Others [52-56] have considered various improvements to this type of global modeling. These models are not directly relevant to the developing fire process because they emphasize the ventilation-controlled fire or they do not consider the local heat transfer process affecting the fire growth. In this regard, Parker and Lee [47] and Fowkes in Ref 11 have attempted a more detailed analysis applicable to the developing fire. Motivated by these approaches, a mathematical model for the developing fire will be presented in the next section.

Before leaving the discussion of mathematical modeling, the consideration of computer solutions should be discussed. This approach solves the governing differential conservation equations by numerical methods. It has been restricted up to now to two-dimensional (or axisymmetric) geometries. This approach was initiated by Torrance and Rockett [57] and has been followed by others [58–60] with various degrees of complexity included in the problem. Attempts to include radiant heat transfer, combustion, turbulence, and openings in the enclosure have all been made. As this approach seeks to model more of reality, confidence in its results must be developed. The application of computer modeling will ultimately depend on its accuracy, cost, and accessibility to the designer.

## A Mathematical Model of Fire Growth

A sufficient understanding exists so that a conceptual model of the physical processes of fire spread in a room can be constructed. In addition, a mathematical description of these processes can be represented in a reasonably simple manner. The chemical aspects of a developing room fire involving the efficiency of combustion and possible reignition of unburned combustion products must await future developments and will not be considered here.

The model to be considered here is motivated by the analyses of room fires by Rockett [61], Fowkes [11], and Zukoski [62] and of corridor fire exposures by this author [63] and McCaffrey and Quintiere [64]. The model considers a fire of a fixed circular burning area on the floor of the room. Only a doorway opening will be considered in the example to follow, and the region beyond the doorway is large in extent. Fresh air flows through the lower part of the doorway, is entrained into the fire plume, heated, and flows out of the upper doorway region. The fire plume heats its local surroundings, and a hot, well-stirred region is formed in the upper volume of the room heating the walls and ceiling by convection and radiation and the floor by radiation alone. The fire grows or not, depending on the resultant heat transfer to other combustible items and on the extent of fuel available. In the following analysis, the contributing mechanisms will be decoupled into three processes: (a) fluid flow to the fire and room, (b) heat transfer within the room, and (c) heat feedback to the fuel and its subsequent consequences on burning rate, flame spread speed, and ignition.

*Fluid Flow*

Figure 5 displays the fluid flow and vertical temperature distribution in a scale model of a floor fire in a room adjoining a corridor. The flow

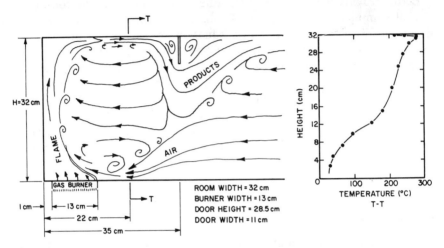

FIG. 5—*Fire induced room flow pattern and vertical temperature distribution taken from a scale model.*

pattern was sketched from titanium oxide ($TiO_2$) smoke traces. Although this is a scale model, it would agree qualitatively with the flow in a full-scale room. The mathematical model that will be adopted follows that of Rockett [61]. The model is illustrated by Fig. 6. A combustion region is depicted above the fire base area $A_v$. This fire plume entrains air up to height $X_d$ (the thermal discontinuity) above which the combustion products are at a uniform temperature $T_g$. This representation is reasonable since, except for possible mixing at the doorway, air can only flow into the upper room atmosphere through the fire plume. Also, except for regions of combustion, the upper atmosphere fluid is well mixed by the entrainment process shown in Fig. 5 so that a uniform temperature $T_g$ is a good first approximation. The flow into and out of the doorway is governed by the height $X_d$ and the neutral plane height $X_n$. (The neutral plane height is the height at which the pressure difference across the doorway is zero.) The doorway flow model has been recently analyzed by Prahl and Emmons [65] who give a derivation of these governing equations.

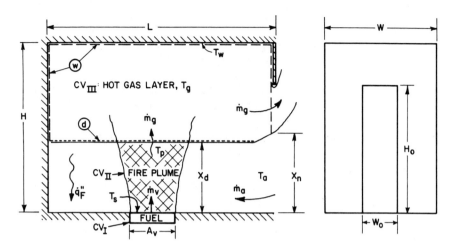

FIG. 6—*Flow and thermal model.*

*Mass Flow Rate of Products Out of the Doorway*

$$\dot{m}_g = 2/3 \; cW_o\varrho_a\sqrt{2gd(1 - d)} \; (H_o - X_n)^{3/2} \tag{5}$$

where

$d$ = density ratio $\varrho_g/\varrho_a$ and
$c$ = orifice coefficient.

*Mass Flow Rate of Air into the Doorway*

$$\dot{m}_a = 2/3 \; cW_o\varrho_a \sqrt{2g(1 - d)} \; (X_n - X_d)^{1/2} \; (X_n + X_d/2) \tag{6}$$

The remaining equation that is required to complete the flow model is the rate of air entrained into the fire plume. For the axisymmetric fire plume an approximation of Steward's [66] result was used.

*Mass Flow Rate of Air Entrained into the Fire Plume*

$$\dot{m}_a = \dot{m}_v\omega(\beta X_d/H_o + 1)^{5/2} \tag{7a}$$

where $\omega$ and $\beta$ are dimensionless parameters. They are given as

$$\omega = \cfrac{1}{1 + \cfrac{\Delta H}{rC_gT_a}} \tag{7b}$$

which is small (to be taken as 0.08 in the calculations to follow), and

$$\beta = 4/5 \, k_e^{4/5} \, (1 - \omega) \left[ \frac{5g\pi^2\varrho_a^2}{12\omega^3\dot{m}_v^2} \right]^{1/5} H_o \qquad (7c)$$

where $k_e \sim 0.1$ is the plume entrainment constant. The weak link in the flow model appears to be the entrainment Eq 7 for several reasons. First, in the form given an uncertainty in the entrainment constant affects $\dot{m}_a$ by almost $k_e^2$. Second, the equation provides a model of a free burning fire, not one that is in an enclosure. Third, the shape of the fire source (that is, axisymmetric, line fire, wall fire, etc.) and drafts in the room which affect the orientation of the flame require a more complete mathematical representation than that given by Eq 7. Nevertheless, the present representation will be qualitatively correct and can be made more accurate by an empirical adjustment in $k_e$. It should be noted that the effect of fire location in the room can be achieved by reducing $k_e$ by 1/2 for a fire against a wall and by 1/4 for a corner fire. Equations 5, 6, and 7 can now be solved for the three unknowns $X_n$, $X_d$, and $\dot{m}_a$, provided it is recognized that the conservation of mass

$$\dot{m}_g = \dot{m}_v + \dot{m}_a \qquad (8)$$

holds and that for a density ratio ($d$) below 1/2, the temperature has little effect on the flow. The flow equations will be decoupled from the energy equation to follow by specifying $d = 1/2$. The fuel mass generation rate, $\dot{m}_v$, will be specified, but the fire area, $A_v$, will be determined to match the energy requirements.

Some results from this flow model are shown in Figs. 7 and 8. The equations were solved by the Newton-Raphson method. Figure 7 displays $\dot{m}_a$ as a function of the traditionally used "ventilation parameter" $A_o\sqrt{H_o}$. The results are in qualitative agreement with the limiting form

$$\dot{m}_a = k_o A_o \sqrt{H_o} \qquad (9)$$

where Thomas et al [51] gives $k_o = 0.50$ kg/m$^{5/2}\cdot$s for the small opening and $k_o \sim 0.13$ for a large opening comparable to the cross section of the compartment. It should be emphasized here that, when the opening is nearly equal to the cross section of the compartment, then Eq 5 and 6 should be  modified to account for small changes in velocity across the opening, and values for the orifice coefficients will be larger than the commonly used values of 0.6 to 0.7. Figure 8 shows that the neutral plane height decreases as the door width decreases and $\dot{m}_v$ increases. The thermal discontinuity $X_d$ is only a few centimeters below $X_n$ in these calculations. These conditions are likely to hold up until flashover, after which the plume entrainment model may not hold in its present form.

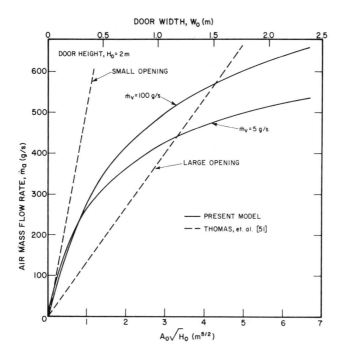

FIG. 7—*Theoretical airflow rate as a function of the ventilation factor,* $A_0\sqrt{H_0}$. *Door height,* $H_0 = 2\ m$; $k_e = 0.1$; $d = 0.5$.

## Energy Transport

In order to keep the present analysis brief and simple to follow, all the thermal processes will be assumed as quasi-steady. Primarily, the transient conduction processes and the phenomenon of transient fire spread, which are the important transient aspects of the problem, will be taken as independent of time. Consequently, the results of this analysis should be interpreted as follows. For a given fire size $A_v$, the mathematical model predicts the flow, thermal, and burning conditions. Based on those results, the potential for fire growth or flashover can be assessed. In addition to the quasi-steady specification, the following assumptions are made.

1. All combustion occurs in the plume below $X_d$. (This is not true in reality since flames can be observed above $X_d$ in room fires and may impinge on and stretch across the ceiling. It is imposed to simplify the problems. A more general model can not be developed at present because the mixing process which controls the rate of combustion in the plume is not well understood.)

2. The upper fluid region is a grey gas at emissivity $\varepsilon_g$ and uniform temperature $T_g$. (This assumption will be satisfactory provided no combustion occurs in this upper region. The recirculatory flow in this region

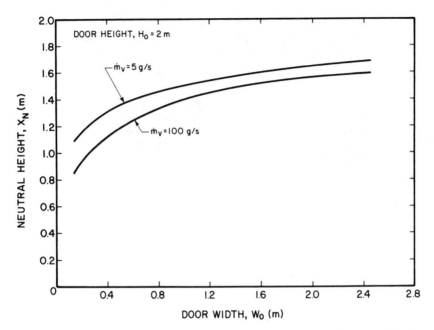

FIG. 8—*Theoretical doorway neutral plane height as a function of door width. Door height,* $H_0 = 2 m$; $k_e = 0.1$; $d = 0.5$.

will tend to make the temperature and composition uniform. However, if combustion occurs here, a uniform temperature would not occur.)

3. All surfaces are blackbodies. (In most cases, the surfaces of materials would have emissivities and absorptivities of 0.8 or higher. However, polished metal surfaces will have high reflectivities, and white painted surfaces would have a high reflectivity for radiation in the visible wavelength region of the spectrum. All surfaces will become more absorbing as soot is deposited on them or as surface charring occurs due to the fire development.)

4. Radiation loss through the doorway is neglected. (In terms of an energy balance, an insignificant fraction of the total energy released in the fire is radiated through openings in the room. Of course, this depends on the size and location of the openings. However, for a single doorway opening, Croce [10,11] found that, during the developing stage through flashover, less than 10 percent of the energy release rate was radiated through the doorway. Although it is justifiable to neglect this quantity in an overall energy balance, the resultant radiant heat flux to exterior surfaces in the field of view for the opening may not be insignificant in terms of potential ignition and flame spread to these surfaces.)

5. Radiant energy transfer from the fire plume is approximated with a constant flame temperature, $T_f$, and is only allowed to occur to the fuel

surface at $T_s$ and the cold floor at $T_a$. However, incident radiant heat flux ($\dot{q}_{r,d}''$) is allowed to strike the fire plume and interact with it and the fuel surface. (This assumption is consistent with Assumption 1 and simplifies the radiative heat transfer analysis by not allowing an interchange between the fire plume and the hot upper gas region. This omission is compensated by the transport of convected energy from the fire plume to the hot gas region.)

6. The floor is at the inlet air temperature $T_a$. (This is valid for the early stage of the fire only. As the floor increases in temperature, the net radiative heat flux to the floor will decrease as significant radiation is emitted from the floor. In the present analysis, holding the floor temperature fixed at $T_a$ yields, more correctly, the incident flux to the floor rather than the net radiative heat flux.)

The energy conservation law will be applied to three spatial regions (or control volumes, CV) which designate distinct regions and which simply characterize the phenomena. These are shown in Fig. 6 and designated as follows.

$CV_I$—The region enclosing a fixed mass of fuel bounded by its exposed surface and an internal horizontal plane which moves at the regression speed of the vaporizing surface and where the fuel is at its initial temperature, $T_a$.

$CV_{II}$—The region enclosing the fire plume bounded by the fuel surface and thermal discontinuity plane at $X_d$.

$CV_{III}$—The region bounded by the walls, ceiling, and thermal discontinuity plane $X_d$.

*Enthalpy Relationships*—Before developing the energy conservation relationships, some thermodynamic state equations will be specified. Primarily, the equations expressing enthalpy ($i$) as a function of temperature ($T$) are needed for the fuel, air, and products. A reference state at $T_a$ (and normal atmospheric pressure) is used. The enthalpy of formation is designated as $i°$, and is zero for air. The fuel is assumed to vaporize from the liquid (or solid) phase as a pure substance. Its enthalpy can be expressed as

$$i_{\text{fuel}} = i_{\text{fuel}}° + C_{\text{fuel}}(T - T_a) \text{ for } T < T_s \qquad (10a)$$

and

$$i_{\text{fuel}} = i_{\text{fuel}}° + C_{\text{fuel}}(T_s - T_a) + \Delta H_v + C_g(T - T_s) \text{ for } T > T_s \qquad (10b)$$

where

$T_s$ = vaporization temperature and
$\Delta H_v$ = heat of vaporization.

The specific heat of the fuel (condensed phase), $C_{fuel}$, is constant, and the gaseous specific heat, $C_g$, will be taken as the same for the air, fuel, and products. The enthalpy relationships for the products (all gaseous) and air follow

$$i_{prod} = i_{prod}{}^\circ + C_g(T - T_a) \tag{11}$$

and

$$i_a = C_g(T - T_a) \tag{12}$$

*Energy Conservation for the Fire Plume*—The application of a steady-state conservation of energy for the fire plume ($CV_{II}$) is given by

$$\varepsilon_f \dot{q}_{r,d}{}''A_v - \dot{q}_s - \dot{q}_{r,f} = \dot{m}_g i_{out}(T_p) - \dot{m}_a i(T_a) - \dot{m}_v i_{fuel\,(gas)}(T_s) \tag{13}$$

The first heat transfer term represents the rate of radiant energy received from the upper hot region of the compartment which is absorbed by the fire plume. As a simplication, the area associated with this transfer of energy was taken as the fuel vaporization area $A_v$. The fire emissivity $\varepsilon_f$ (or absorptivity) is represented by

$$\varepsilon_f = 1 - \exp(-k_f H_f) \tag{14}$$

where $k_f = 0.5 \text{ m}^{-1}$ is an approximate value for wood flames [67] assuming their height ($H_f$) is about twice their base diameter. The height of a turbulent diffusion flame can be determined by

$$H_f = 16.2 \left[ \frac{(r + \omega)^2 \omega}{\varrho_a{}^2 g(1 - \omega)^5} \right]^{1/5} \dot{m}_b{}^{2/5} \tag{15}$$

from Steward's correlation [66]. For a fire plume in a room, this equation would only apply provided $H_f < X_d$. Hence, in the present model, $H_f$ in Eq 14 should be the smaller of $X_d$ or that given by Eq 15.

The other two heat transfer terms represent the net heat transfer rate to the fuel surface ($\dot{q}_s$), and the rate of radiative loss from the fire plume ($\dot{q}_{r,f}$). The former will be explicitly expressed later. The radiative heat flux from the plume, neglecting an interchange to the upper surface of $CV_{II}$ (Assumption 5) can be determined by using standard principles [68] as

$$\dot{q}_{r,f}{}'' = \varepsilon_f \sigma T_f{}^4 + F_{23}(1 - \varepsilon_f)\sigma T_s{}^4 - [1 - F_{22}(1 - \varepsilon_f)]\sigma T_a{}^4 \tag{16a}$$

where $F_{22}$ and $F_{23}$ are appropriate configuration factors. Because $T_f$ is large ($\sim$1200 K) compared to $T_s$ and $T_a$, this heat flux was approximated as

$$\dot{q}_{r,f}'' \cong \varepsilon_f \sigma T_f^4 \qquad (16b)$$

Finally, by approximating the fire plume as a cylinder of base area $A_v$ and height $H_f$

$$\dot{q}_{r,f} = \varepsilon_f \sigma T_f^4 2\sqrt{\pi A_v} H_f \qquad (17)$$

Although this approximation is crude, it probably would not be fruitful to make it more sophisticated until our knowledge of radiation from fires has improved [67,69,70].

For the enthalpy flow rate terms in Eq 13, only $i_{out}(T_p)$ must be determined. This can be done by determining the mass fractions of species based on the combustion reaction occurring in the plume. The reaction can be represented on a mass basis as

$$1[\text{fuel}] + r[\text{air}] \rightarrow (1 + r)\,[\text{products}]$$

where $r$ is the stoichiometric mass air to fuel ratio. Depending on the fuel vaporized ($\dot{m}_v$) and air entrained into the plume ($\dot{m}_a$), there may be excess fuel ($\dot{m}_{py}$) or excess air leaving the plume. In general

$$\dot{m}_v = \dot{m}_b + \dot{m}_{py} \qquad (18)$$

where

$\dot{m}_b$ = fuel burned and
$\dot{m}_{py}$ = 0 if $\dot{m}_v \leqslant \dot{m}_a/r$.

By defining the heat of combustion in the conventional manner as

$$\Delta H = i_{\text{fuel}}{}^\circ - (1 + r)i_{\text{prod}}{}^\circ \qquad (19)$$

that is, based on the condensed fuel phase, then it can be shown that Eq 13 can be written as

$$\varepsilon_f \dot{q}_{r,d}'' A_v + \dot{m}_b \Delta H - \dot{q}_s - \varepsilon_f \tau T_f^4 \times 2H_f \sqrt{\pi A_v} = \dot{m}_g C_g (T_p - T_a) \atop + \dot{m}_{py}[\Delta H_v + (C_{\text{fuel}} - C_g)(T_s - T_a)] - \dot{m}_v[\Delta H_v + C_{\text{fuel}}(T_s - T_a)] \qquad (20)$$

*Energy Conservation for the Fuel*—By assuming an inexhaustible supply of fuel and no heat losses from the fuel except at its vaporizing surface, the conservation of energy for $CV_1$ yields

$$\dot{q}_s = \dot{m}_v[\Delta H_v + C_{fuel}(T_s - T_a)] \tag{21}$$

This equation can now be used to simplify Eq 20 by cancelling the heat supply rate to the fuel with the rate of energy necessary to produce gaseous fuel.

*Energy Conservation for the Hot Gas Layer*—An energy conservation applied to the hot gas region at $T_g$ ignoring any radiative interaction with the fire yields

$$\dot{m}_g C_g(T_p - T_g) = \dot{q}_{r,w}''A_w + \dot{q}_{r,d}''A_F + h_w A_w(T_g - T_w) \tag{22}$$

where $\dot{q}_{r,w}''$ and $\dot{q}_{r,d}''$ are the net radiative heat fluxes to the hot walls and ceiling at $T_w$ and through the thermal interface (labeled $d$ in Fig. 6), respectively. The hot wall and ceiling area is given as

$$A_w = LW + 2[(H - X_d)(W + L)] \tag{23}$$

and the floor area as

$$A_F = LW \tag{24}$$

The radiative heat fluxes are determined from interchange relationships between the hot gas and surfaces $W$ and $d$ [66]

$$\dot{q}_{r,w}'' = \varepsilon_g \sigma T_g^4 + \gamma(1 - \varepsilon_g)\sigma T_a^4 - [1 - (1 - \gamma)(1 - \varepsilon_g)]\sigma T_w^4 \tag{25}$$

$$\dot{q}_{r,d}'' = (1 - \varepsilon_g)\sigma T_w^4 + \varepsilon_g \sigma T_g^4 - \sigma T_a^4 \tag{26}$$

The radiative heat flux to the floor due to the upper hot region in the room is given as

$$\dot{q}_F'' = F_{dF}\dot{q}_{r,d}'' \tag{27}$$

where the geometric configuration factor $F_{dF}$ is for parallel planes and depends on $X_d$, $L$, and $W$ [71]. This does not include direct radiation from the fire plume.

The heat transfer coefficient, $h_w$, and gas emissivity, $\varepsilon_g$, depends on other variables, and relationships are needed to specify their values. Alpert [72] and Veldman et al [73] have shown that the heat transfer to the ceiling from an impinging buoyant plume depends on the energy release rate of the fire, the height of the ceiling, and the radial distance from the plume. This is an idealized result for the room fire since they do not consider the confining effect of the room and doorway. The effect of a room on ceiling heat transfer has been recently investigated by Zukoski and Kubota [74]. Also, Moriya et al [75, 76] measured an average convective heat transfer coefficient to a cylinder in the center of a fully developed room fire. It is likely that the heat transfer coefficient may vary from 1 to 100 $W/m^2K$ depending upon the fire condition. A $h_w$ of 20 $W/m^2K$ was selected for calculations in the present analysis. For the gas emissivity an approximate relationship will be specified.

$$\varepsilon_g = 1 - \exp[-(k_1 + k_2 C_s)(H - X_d)] \tag{28}$$

where $C_s$ is the soot concentration in mg/litre, $k_1 = 0.33$ m$^{-1}$ (which corresponds to roughly 12 percent water ($H_2O$) and 12 percent carbon dioxide ($CO_2$) in the gas), and $k_2 = 0.47$ litre/mg·m. The values of $k_1$ and $k_2$ were estimated from an analysis of smoke in corridor fires by Bromberg and Quintiere [77]. As seen by this model, radiative effects have been ignored which involve scattering due to liquid aerosols and changes in radiative transport due to temperature and concentration gradients.

*Conservation of Energy at the Heated Walls and Ceiling*—The final equation required to complete the solution for $T_g$, $T_w$, and $\dot{q}_F''$ is the wall surface and gas energy balance. This is given as

$$\dot{q}_{r,w}'' + h_w(T_g - T_w) = K(T_w - T_a) \tag{29}$$

which says that the sum of the net radiative and convective fluxes to the walls and ceiling are equal to the heat conducted into the wall. In the early growth of a fire, the conductive heat is, for typical wall constructions, absorbed by the walls not conducted through them. For this reason, the conductance $K$ is approximated as

$$K = \sqrt{\frac{k\varrho c}{t_b}} \tag{30}$$

where $k\varrho c$ is a thermal property of the wall lining and $t_b$ can be regarded as a characteristic burning time, that is

$$t_b = m_b/\dot{m}_b \tag{31}$$

the duration of the fire. The characteristic conduction time for the wall or the time to heat the unexposed fire side can be given as

$$t_w = \frac{\delta^2}{(k/\varrho c)} \tag{32}$$

where $\delta$ is the thickness of the wall lining. For typically used plaster wall-board, $t_w$ can be roughly 20 min. Values of $K$ can be 10 to 30 W/m²K for plaster board, depending on $t_b$. The range of probable values of $t_b$ is 10 to 100 min, depending upon the amount of fuel available. The effect of $K$ or $k\varrho c$ on the results will be shown later.

### Fire Spread and Growth

The growth and spread of the fire will depend on the heat transfer from the flames and heat feedback from the room environment to the unburnt available fuel. The fire growth process can be separated into five distinct modes:

1. Continuous surface flame spread;
2. Contiguous fire spread, that is, spread to items in contact with the primary fire or adjoining items;
3. Noncontiguous fire spread, that is, spread to items separated by a relatively large distance from the fire;
4. An increase in the burning rate due to room augmented heat transfer to the burning item; and
5. Propagation through a combustible gas mixture which evolved from unburnt pyrolysis products.

The following discussion will elaborate, to a limited extent, on all but the last mode.

The rate of continuous surface flame spread will depend on the point of ignition, surface orientation, and other configuration and thermophysical variables. As an illustrative example, flame spread over a horizontal surface with an externally applied radiant flux is considered. This problem was analyzed by Kashiwagi [78], and his results for flame convective and radiative forward heat fluxes are shown in Fig. 9. He found that both the burning rate and spread rate increase significantly with an increase in the externally applied radiative heat flux. This is demonstrated mathematically by [63]

$$V_f = \frac{4}{\pi} \frac{\delta_f}{(k\varrho c)_s} \left[ \frac{\dot{q}_f''}{(T_{ig} - T_a) - (\dot{q}_F''/h)(1 - e^{\tau} erfc\sqrt{\tau})} \right]^2 \tag{33}$$

where

$\delta_f$ = effective heat transfer length over which an average total heat flux $\dot{q}_f''$ from the flame is incident,

$h$ = total heat transfer coefficient, and

$\tau = th^2/k\varrho c$.

It shows how the velocity $V_f$ can increase rapidly with the external flux

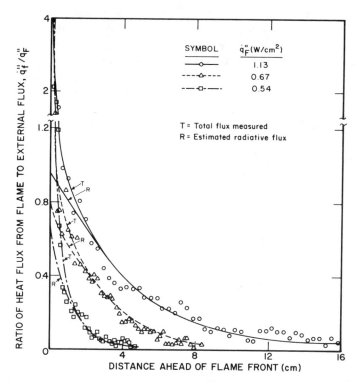

FIG. 9—*Heat flux ahead of a flame moving horizontally over a carpet under various external heat fluxes* [78].

$\dot{q}_F''$ to the floor and time $t$. In theory, as the denominator of Eq 33 approaches zero, a form of flashover might be perceived, provided enough fuel is available.

Continguous spread relies on both convective heating from the flame plume and direct radiation transfer fromthe flame. Of course, it is also enhanced by preheating from an external heat flux resulting from the fire being confined by the room. As an example of quantifying the radiative heat fluxes, the analysis by Dayan and Tien [79] for a cylindrical fire column should be considered. They show that the radiant flux from the flame at large distances, $D$, decreases as $D^{-2}$ for surfaces facing the flame and is proportional to $R_f H_f^2 / D^3$ for horizontal surfaces, where $R_f$ is the flame radius and $H_f$ is its height. A demonstration by de Ris and Orloff [80] of the surface heat flux by flames on a vertical surface showed the significance of the radiative contribution. Their results indicate that convection and radiation are both important in analyzing heat transfer from flames in contact with a surface. A general prediction model for this is not yet available.

Noncontiguous flame spread is likely to occur by two mechanisms. The flame plume impinges on the ceiling and ignites or simply pyrolyzes any available fuel of the ceiling and wall linings, and the pyrolysis products may ignite elsewhere when they encounter sufficient oxygen. The second way is by spontaneous or piloted ignition of the combustibles in the lower space of the room due to radiative heating by the hot upper room region. For example, Simms [81] states that about 4 W/cm² is sufficient to spontaneously ignite cellulosic materials in 30 s or more. The minimum radiative flux required for spontaneous ignition is given as 2 to 4 W/cm² and for piloted ignition as low as 1 W/cm² [82].

Finally, another mode of fire growth is the increase in burning rate per unit area due to an increase in heat flux to the burning surface from the surroundings. Friedman [2], in discussing some recent work, reports as much as a 60 percent increase in the burning rate of wood cribs in an enclosure over the free burning rate for the wood cribs; whereas the free burning rate of a plastic horizontal slab was enhanced by almost three times when burned in 1.22-m-high enclosure. The radiative interactions occurring here are very complex, and only the free burning problem has been mathematically modeled [83]. In order to complete the system of equations considered in the present analysis, such a feedback relationship is needed. This has been represented by Eq 21; however, an expression for $\dot{q}_s$ is needed. Because a fire within an enclosure is being considered, $\dot{q}_s$ is composed of two distinct contributions: the heat transfer from the fire plume and the heat transfer radiated from the hot gas layer and heated enclosure surfaces. By considering the radiative exchanges on the fire plume and by assuming a simple form for convective heat transfer from the fire to the fuel, it follows that

$$\dot{q}_s = A_v[(1 - \varepsilon_f)\dot{q}_{r,a}{}'' + \varepsilon_f \sigma T_f^4 + (1 - \varepsilon_f)\sigma T_a^4 - \sigma T_s^4 + h_s(T_f - T_s)]$$

(34)

The first term in brackets on the right-hand side represents the radiative heat flux which results from the fire being confined within the room. The second and last terms represent the radiative and convective heat transfer from the fire to the fuel, respectively. Although this representation is simplistic, it does qualitatively include the primary features of these heat transfer processes. Equation 34 is the last equation required to complete the necessary number of equations for solving the room fire growth model.

The solution to the presently formulated model is performed as follows. The flow equations (Eqs 5–7) are solved for $\dot{m}_a$, $X_d$, and $X_n$ given $\dot{m}_v$ and the doorway dimensions. Then, the energy transport equations (Eqs 20, 21, 25, 26, 29, 34), including subsidiary relationships, are solved to yield $T_g$, $T_w$, $\dot{q}_{r,a}{}''$, $\dot{q}_{r,w}{}''$, $\dot{q}_s$, and $A_v$. The floor heat flux, $\dot{q}_F{}''$, due to the upper hot gas layer, is then determined from Eq 27. The fire plume also contributes to radiative floor heat transfer and decreases as a function of distance from the fire. For a relatively large distance from the fire, $\dot{q}_F''$ can be used as an indicator of flashover in the model. For example, flashover may be considered to have occurred if $\dot{q}_F'' \geqslant 2$ W/cm$^2$. Alternately, the gas temperature $T_g$ may be used as an indication of flashover or potential ignition of fuel in the upper room space. In summary, the solution of the room fire model yields, for a given room geometry, materials, and fuel: the room gas and ceiling temperature, floor radiative heat flux, and fuel mass loss rate and burning rate.

*Illustrative Results*

In order to obtain numerical results from the model, some geometric and fuel parameters must be specified. These were chosen to represent typical rooms and a hypothetical fuel suggestive of a plastic. Table 1

TABLE 1—*Specified parameters for the fire growth model.*

| | |
|---|---|
| *Room and Doorway Dimensions, m* | |
| $L$, $W$ (length, width) | 3, 3 and 4, 4, |
| $H$ (height) | 2.5 |
| $W_o$ (door width) | 0.25 to 2 |
| $H_o$ (door height) | 2 |
| *Fuel Parameters* | |
| $r$ (air fuel ratio) | 7 kg air/kg fuel |
| $\Delta H$ (heat of combustion) | 20 MJ/kg |
| $T_s$ (pyrolysis temperature) | 600 K |
| $\Delta H_v$ (heat of volatilization) | 1.6 MJ/kg |
| $C_{\text{fuel}}$ (specific heat of condensed phase fuel) | 2 kJ/kg·K |
| *Other Parameters* | |
| $T_f$ (flame temperature) | 1200 K |
| $C_g$ (specific heat of gases) | 1.05 kJ/kg·K |
| $T_a$ (air temperature) | 300 K |
| $h_w$ (wall convective coefficient) | 20 W/m$^2$·K |
| $K$ (wall conductance) | 10, 30 W/m$^2$·K |
| $C_s$ (soot concentration) | 0.05, 4 mg/litre |
| $h_s$ (fuel convective coefficient) | 5 W/m$^2$·K |

summarizes the values selected for the computations that were made. The equations were solved on a digital computer for fuel mass loss rates, $\dot{m}_v$, between 5 to 100 g/s in increments of 5 g/s. As stated earlier, the area involved in combustion is treated as the independent variable and represents the fire size.

Figure 10 displays results for the burning rate or, more specifically, the fuel volatilization rate $\dot{m}_v$ for a wall material of conductance ($K$), room size ($L$, $W$), and smoke concentration ($C_s$). Above 61 g/s, the maximum combustion rate is reached, and further increases in area $A_v$

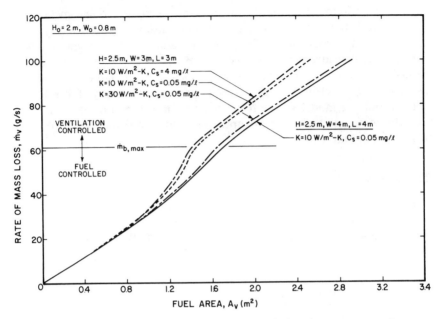

FIG. 10—*Theoretical mass loss rate as a function of fuel surface area, room size, conductance (K), and smoke concentration ($C_s$).*

lead to unburned products in the room. Thus, the extent of flames at the doorway will increase significantly. A change in $C_s$ from 0.05 to 4 mg/litre represents a change from very little smoke to almost an opaque black cloud respectively; whereas, values of $K$ of 10 and 30 W/m²K represent plaster board walls and concrete respectively. The results in Fig. 10 show $C_s$ has a small effect; yet a change in $K$ from 10 to 30 W/m²K yields almost the same decrease in $m_v$ as a change in room size from 3 by 3 m to 4 by 4 m. The reason why $C_s$ has little effect on $\dot{m}_v$ (and also on $\dot{q}_F''$) is that the corresponding change in $\varepsilon_g$ from 0.3 to about 1 causes the gas to lose more heat to walls by radiation, which increases $T_w$ and decreases $T_g$.

If the free burning fire (that is, no enclosure) is analyzed by the present model, it will be found that the model yields no increase in the burning rate within an enclosure compared to the free burning fire. This is contrary to results observed in experiments [2]. Consequently, this must be viewed as a deficiency in the present model and, to a lesser extent, may be attributed to the choice of parameters assumed. Moreover, the model for radiative heat transfer to the fuel probably needs refinement.

However, the present model does show the relative influence of room size and materials on the fuel mass loss rate. Figure 11 shows the effect of these parameters on the incident radiative heat flux to the floor, $\dot{q}_F''$. If

Monitoring of streams may be by the observation of change in the major components of the ecosystem in selected stream areas or by the introduction of substrates to study one or more groups of organisms. Studies of whole ecosystems are much more informative and much more expensive. The monitoring by the use of substrates is very inexpensive and is feasible on a continuing basis.

Many new methods of monitoring the biosphere will have to be devised through research. At the present time in the aquatic world we probably have the best examples of fairly long-term monitoring, that is, continual monitoring and intermittent monitoring over a 20 to 25 year period.

## Background of Monitoring Sites

I have selected two examples of these types of monitoring, that is, monitoring using natural communities of aquatic organisms as a basis on which to measure change. The two examples that I have selected are a free-flowing river in the southeastern part of the United States, the Savannah River (which is the boundary between South Carolina and Georgia) and the Sabine River estuary in Texas.

### Savannah River

The first example, the Savannah River, has had for many years a fair amount of sanitary wastes, farm wastes, and sediments entering it, but at the time when our studies were started these had an effect but had not degraded the river severely. Indeed, during the course of our studies this river has not been severely degraded, but we have been able to see the effects of various types of man's activities upon it. Until the middle 1960's, there were very little industrial wastes entering the river between Augusta and our study area at the Savannah River Plant. However, during the 1960's, several industries were constructed between Augusta and the study area. A large dam was put into operation on the river upstream from Augusta between 1951 and 1955, and dredging was carried out to deepen the channel between 1958 and 1962. Industrialization developed significantly in the late 1960's.

### Sabine River Estuary

The second example, the Sabine River estuary, is a river which had a heavy load of organic pollution at the time our studies started. Since the start of our study, waste treatment facilities have been greatly enlarged and improved. At the same time, many more industries have come into the area, and the population of the city or Orange and the surrounding area have increased markedly. A dam had been built during this period of our study (Fig. 1).

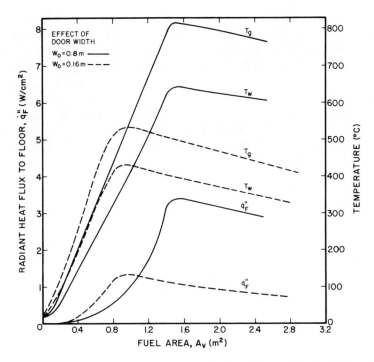

FIG. 12—*Theoretical thermal conditions as a function of room fire size* ($A_v$) *and door width* ($W_o$). *Room size: 3 by 3 by 2.5 m high;* $H_o = 2$ m; $C_s = 0.05$ mg/litre; $K = 10$ $W/m^2 \cdot K$.

the ventilation parameter, $A_o\sqrt{H_o}$, in Fig. 13. There, the available fuel area $A_v$ is analogous to the crib weight of Fig. 1. The locus of $T_g = 600°C$ and $\dot{q}_F'' = 2$ W/cm² are shown, and the regions contained by these loci can be considered as flashover regions similar to that of Fig. 1. Thus, the mathematical results are similar to these data. It should be noted, however, that the theoretical ventilation-controlled mass loss rates do not coalesce into one line as do the experimental wood crib results of Fig. 1. In Fig. 13, the ventilation-controlled fires are mass loss rates above the stoichiometric locus curve. The region below this locus is the region of fuel-controlled fires.

## Conclusions

There is an abundance of experimental results for fire development in rooms involving furnishings. These suggest that, in typical residential size rooms, flashover is likely to occur if the major item of furniture becomes fully involved in fire, or if something like a large waste container or chair fire ignites a combustible wall and ceiling lining. These experiments have main characteristics of water quality. Sediment load, bacterial content, temperature, and flow rates are also estimated. The sediments that form the river bed are also analyzed for particle size and source of possible pollutants such as heavy metals and petroleum. Such studies are usually performed on both high and low tides in estuaries, and at different periods of the day in free-flowing streams. Sufficient analyses are made on different days at different times to obtain a mean and a standard error of the mean for the various analyses and a confidence level of predictability.

In the biological studies, the specialist enters the stream and thoroughly collects the area. He usually continues to collect until he has collected for half an hour without adding a new species to his collection. This, of course, is true for the macroscopic species; for microscopic organisms, it is important to collect all habitats in the study reach. Typically it takes about two days for a specialist to study an area of 200 to 300 yd in length. It is important that all types of habitats characteristic of that particular reach of the river are included in the study area and that they be of comparable types in the various areas of study.

## Discussion of Results

### Savannah River

The results of such thorough studies in the Savannah River indicate that there was a considerable improvement in the algae from 1954 to 1955. This is probably due to the operation of the Clark Hill Dam. This dam brought about the precipitation of the sediments, and it was evident that the sediment load of the river downstream was much lower immediately after the operation of Clark Hill Dam. This, in turn, would increase the depth of the photosynthetic zone and thus bring about a production of a larger number of species of benthic organisms. This same effect of improvement, probably due to the improvement of the food base, is seen in algae, arthropods, and fish (Figs. 2 to 4).

In the period around 1960, there was a decided drop in species diversity, particularly of the algae and insects. It was during this time that considerable dredging of the river was performed. As a result, the water had a much higher suspended solids load, the algae decreased, and the food for insects similarly decreased. Also, the filter feeders were found to be much fewer during the period when the river was being dredged. This period extended from around 1960 to 1965. However, toward the end of the dredging period, there was a decided increase in the benthic algae at most of the stations. The only station where this was not true was at Station 6. The fish did not show the decrease in species in 1960 as was indicated by the other groups, but rather they tended to increase a bit over 1955. The decrease in fish species, probably due to a lower food base, was evidenced in 1965.

In 1968, we saw a decided decrease at all stations in the number of algal be done. In this regard, more work is called for in studying the entrainment rates of fires in enclosures, the mixing process which control the rate of combustion, flame height and flame extension length under a ceiling, and the radiative interaction between flames and smoke and its effect on the surfaces and subsequent burning and pyrolysis rates.

## Acknowledgments

The author is indebted to fruitful discussions with Dr. John Rockett and is grateful for the assistance of Karen Den Braven with the computer calculations of the model.

## References

[1] Thomas, P. H., *Fire Research and Development News,* Vol. 1, 1975, pp. 1–7.
[2] Friedman, R., "Behavior of Fires in Compartments," presented at International Symposium Fire Safety of Combustible Materials, Edinburgh, Scotland, 15–17 Oct. 1975, pp. 100–113.
[3] McCarter, R. J., "Polyurethane Smolders?," (to be submitted to *Journal of Fire and Flammability*).
[4] Hafer, C. A. and Yuill, C. H., "Characterization of Bedding and Upholstery Fires," Final Report SwRI Project No. 3-2610, Southwest Research Institute, San Antonio, Tex., 31 March 1970.
[5] Woolley, W. D. and Ames, S. A., "The Explosion Risk of Stored Foamed Rubber," CP 36/75, Building Research Establishment, Watford, England, 1975.
[6] Rodak, S. and Ingberg, S. H., "Full-Scale Residential Occupancy Fire Tests of 1939," NBS Report 9527, National Bureau of Standards, 15 May 1967.
[7] Bruce, H. D., "Experimental Dwelling-Room Fires," No. D1941, Forest Products Laboratory, U.S. Dept. of Agriculture, Madison, Wis., June 1953.
[8] Waterman, T. E., "Determination of Fire Conditions Supporting Room Flashover," Final Report IITRI Project M6131, DASA 1886, Defense Atomic Support Agency, Washington, D.C., Sept. 1966.
[9] Hillenbrand, L. J. and Wray, J. A., "Full-Scale Fire Program to Evaluate New Furnishings and Textile Materials Utilized by the National Aeronautics and Space Administration," Contract No. NASW-1948, Battelle Columbus Laboratories, Columbus, Ohio, July 1973.
[10] Croce, P. A. and Emmons, A. W., "The Large-Scale Bedroom Fire Test," 11 July 1973, FMRC Ser. No. 21011.4, Factory Mutual Research, Norwood, Mass., July 1974.
[11] Croce, P. A., "A Study of Room Fire Development: The Second Full-Scale Bedroom Fire Test of the Home Fire Project (24 July 1974)," Factory Mutual Research, Norwood, Mass., June 1975.
[12] Hägglund, B., Jansson, R. and Onnermark, B., "Fire Development in Residential Rooms after Ignition from Nuclear Explosions," FOA Rep. C 20016-D6 (A3), Försvarets Forskningsanstalt, Stockholm, Sweden, Nov. 1974.
[13] Gross, D. and Fang, J. B., "The Definition of a Low Intensity Fire," NBS Special Publication 361, Vol. 1, National Bureau of Standards, Washington, D.C., March 1972.
[14] Fang, J. B., "Measurement of the Behavior of Incidental Fires in a Compartment," NBSIR 75-679, National Bureau of Standards, Washington, D.C., March 1975.
[15] Theobald, C. R., "The Critical Distance for Ignition from Some Items of Furniture," FR Note 736, Fire Research Station, Borehamwood, England, Dec. 1968.
[16] Christian, W. J. and Waterman, T. E., *Fire Technology,* Vol. 7, No. 3, Aug. 1971, pp. 205–217.
[17] Fang, J. B., "Fire Buildup in a Room and the Role of Interior Finish Materials," NBS Technical Note 879, National Bureau of Standards, Washington, D.C., June 1975.

[18] Castino, G. T., Beyreis, J. R. and Metes, W. S., "Flammability Studies of Cellular Plastics and Other Building Materials used for Interior Finishes," Subject 723, Underwriters Lab. Inc., Northbrook, Ill., 13 June 1975.

[19] Davis, S. and Tu, K. M., "Flame Spread of Carpets Involved in Room Fires," NBSIR 76-1013, National Bureau of Standards, Washington, D.C., June 1976.

[20] Hinkley, P. L., Wraight, H., and Theobald, C. R., "The Contribution of Flames Under Ceilings to Fire Spread in Compartments. Part I. Incombustible Ceilings," FR Note 712, Fire Research Station, Borehamwood, England, 1968.

[21] Hinkley, P. L. and Wraight, H., "The Contribution of Flames Under Ceilings to Fire Spread in Compartments. Part. II. Combustible Ceiling Linings," FR Note 743, Fire Research Station, Borehamwood, England, 1969.

[22] Thomas, P. H., "The Role of Flammable Linings in Fire Spread," Board Manufacture, Sept. 1969.

[23] Quintiere, J., "Some Observations on Building Corridor Fires," Fifteenth Symposium (International) on Combustion, The Combustion Institute, Aug. 1974, pp. 163-174.

[24] McGuire, J. H., Fire Technology, Vol. 4, No. 2, 1968, pp. 103-108.

[25] Christian, W. J. and Waterman, T. E., Fire Technology, Vol. 6, No. 3, Aug. 1970, pp. 165-188.

[26] Christensen, G., Lohse, V., and Malmstedt, K., "Full Scale Fire Tests —The Spread of Fire from a Chamber to a Corridor," SBI Report No. 592, The Danish National Institute of Building Research, Copenhagen, Denmark, 1967.

[27] Rhodes, A. C. and Smith, P. B. in The Use of Models in Fire Research, W. G. Berl, Ed., National Academy Science–National Research Council, Washington, D.C., 1961, pp. 235-255.

[28] Roberts, A. F. and Clough, G., Combustion and Flame, Vol. 11, No. 5, Oct. 1967, pp. 365-376.

[29] de Ris, J., Combustion Science Technology, Vol. 2, 1970, pp. 239-258.

[30] Yokoi, S., "Study on the Prevention of Fire-Spread Caused by Hot Upward Current," Report of Building Research Institute No. 34, The Building Research Institute Japan, Nov. 1960.

[31] Butler, C. P., Martin, S. B., and Wiersma, S. J., "Measurements of the Dynamics of Structural Fires," Fourteenth Symposium, (International) on Combustion, The Combustion Institute, 1973, pp. 1053-1059.

[32] Shorter, G. W. in The Use of Models in Fire Research, W. G. Berl, Ed., National Academy Science—National Research Council, Washington, D.C., 1961, pp. 289-295.

[33] Gross, D. and Breese, J. N., "Experimental Fires in Enclosures" (Growth to Flashover) CIB Cooperative Program (CIB 2), NBS Report 10471, National Bureau of Standards, Washington, D.C., 15 July 1971.

[34] Heselden, A. J. M. and Melinek, S. J., "The Early Stages of Fire Growth in a Compartment," FR Note No. 1029, Fire Research Station, Borehamwood, England, Jan. 1975.

[35] Spalding, D. B., "The Art of Partial Modeling," Ninth Symposium (International) on Combustion, The Combustion Institute, 1963, pp. 833-843.

[36] Hottel, H. C. in The Use of Models in Fire Research, W. G. Berl, Ed., National Academy Science—National Research Council, Washington, D.C., 1961, pp. 32-47.

[37] Williams, F. A., Fire Research Abstract Review, Vol. 11, 1969, pp. 1-23.

[38] de Ris, J., Kanury, A. M. and Yuen, M. C., "Pressure Modeling of Fires," Fourteenth Symposium (International) on Combustion, The Combustion Institute, 1973, pp. 1033-1044.

[39] Kanury, A. M., "Pressure Modeling of Fires III. Pan Fires with a Variety of Plastics," FMRC Series No. 19721-9, Factory Mutual Research, Norwood, Mass., Jan. 1973.

[40] Alpert, R. L., "Pressure Modeling Transient Crib Fires," Paper 75-HI-6, American Society of Mechanical Engineers, San Francisco, Calif., Aug. 1975.

[41] Waterman, T. E., "Scaling of Fire Conditions Supporting Room Flashover," DASA Rep. No. 2031, Defense Atomic Support Agency, Washington, D.C., Dec. 1967.

[42] Waterman, T. E., "Scaled Room Flashover," Final Report OCD WU 2534G, Office of Civil Defense, Washington, D.C., April 1971.

[43] Heskestad, G., "Modeling of Enclosure Fires," Fourteenth Symposium (International) on Combustion, The Combustion Institute, 1973, pp. 1021-1030.

[44] Heskestad, G., "Similarity Relationship for the Initial Convective Flow Generated by Fire," Paper No. 72-WA/HT-17, American Society of Mechanical Engineers, New York, Nov. 1972.

[45] Croce, P. A., "Modeling of Vented Enclosure Fires," Presented at Fall Meeting, Eastern Section, Combustion Institute, APL, Nov. 1974.

[46] Parker, W. J. and Lee, B. T., "Fire Build-up in Reduced Size Enclosure," NBS Special Publication 411, National Bureau of Standards, Washington, D.C., Nov. 1974, pp. 139-153.

[47] Parker, W. J. and Lee, B. T., "A Small-Scale Enclosure for Characterizing the Fire Buildup Potential in a Room," NBSIR 75-710, National Bureau of Standards, Washington, D.C., June 1975.

[48] Hird, D. and Fischl, C. F., Quarterly of the NFPA, National Fire Prevention Association, Oct. 1954, pp. 123-131.

[49] Kawagoe, K., "Fire Behavior in Rooms," Report No. 27, The Building Research Institute, Japan, Sept. 1958.

[50] Sekine, T., "Room Temperature in Fire of a Fire-Resistive Room," Report No. 29, The Building Research Institute, Japan, March 1959.

[51] Thomas, P. H., Heselden, A. J. M. and Law, M., "Fully-Developed Compartment Fires—Two Kinds of Behavior," F. R. Technical Paper No. 18, Fire Research Station, Borehamwood, England, Oct. 1967.

[52] Tsuchiya, Y. and Sumi, K., Combustion and Flame, Vol. 16, No. 2, 1971, pp. 131-139.

[53] Magnusson, S. E. and Thelandersson, S., "Temperature-Time Curves of Complete Process of Fire Development," Acta Polytechnica Scandinavica, Civil Engineering and Building Construction Series, No. 65, Stockholm, Sweden, 1970.

[54] Harmathy, T. Z., Fire Technology, Vol. 8, No. 3, Aug. 1972, pp. 196-217.

[55] Harmathy, T. Z., Fire Technology, Vol. 8, No. 4, Nov. 1972, pp. 326-351.

[56] Thomas, P. H., Fire Technology, Vol. 11, No. 1, Feb. 1975, pp. 42-47.

[57] Torrance, K. E. and Rockett, J. A., Journal of Fluid Mechanics, Vol. 36, No. 1, 1969, pp. 33-54.

[58] Trent, D. S. and Welty, J. R., "Numerical Computation of Convection Currents Induced by Fires and Other Sources of Buoyancy," WSCI 72-18 Battelle, Pacific Northwest Laboratory, Richland, Wash., 1972.

[59] Larson, D. W., "Analytical Study of Heat Transfer in an Enclosure with Flames," doctoral dissertation, Purdue University, West Lafayette, Ind., 1972.

[60] Lloyd, J. R. and Doria, M. L., "Fire and Smoke Spread," Semi-Annual Progress Report, University of Notre Dame, South Bend, Ind., May 1975.

[61] Rockett, J. A., "Fire Induced Gas Flow in an Enclosure," accepted for publication, Combustion Science and Technology.

[62] Zukoski, E. E., "Convective Flows Associated with Room Fires," Semi-Annual Progress Report NSF GI 31892XI, California Institute of Technology, Pasadena, Calif., June 1975.

[63] Quintiere J., "The Application and Interpretation of a Test Method to Determine the Hazard of Floor Covering Fire Spread in Building Corridors," International Symposium Fire Safety of Combustible Materials, Edinburgh, Scotland, 15-17 Oct. 1975, pp. 355-366a.

[64] McCaffrey, B. J. and Quintiere, J. G., "Fire Induced Flow in a Scale Model Study," CIB Symposium on the Control of Smoke Movement in Building Fires," Grarston, England, 4-5 Nov. 1975, pp. 34-47.

[65] Prahl, J. and Emmons, H. W., Combustion and Flame, Vol. 25, No. 3, Dec. 1975, pp. 369-385.

[66] Steward, F. R., Combustion Science and Technology, Vol. 2, 1970, pp. 369-385.

[67] Hägglund, B. and Persson, L. E., "An Experimental Study of the Radiation from Wood Flames," FOA 4 Rapport, C 4589-D6(A3), Försvarets Forskningsanstalt, Stockholm, Sweden, June 1974.

[68] Siegel, R. and Howell, J. R., Thermal Radiation Heat Transfer, Vol. III, NASA SP-164, National Aeronautics and Space Administration, Washington, D.C., 1971.

[69] Yumoto, T., "An Experimental Study on Heat Radiation from Oil Tank Fire," Report No. 33, Fire Research Institute, Japan, 1971.

[70] Markstein, G. H., "Radiative Energy Transfer from Turburlent Diffusion Flames," Paper No. 75-HT-7, American Society of Mechanical Engineers, San Francisco, Calif., Aug. 1975.

[71] Hamilton, D. C. and Morgan, W. R., "Radiant-Interchange Configuration Factors," NACA TN 2836, National Advisory Committee on Aeronautics, Washington, D.C., Dec. 1952.

[72] Alpert, R. L., "Fire Induced Turbulent Ceiling-Jet, FMRC Serial No. 19722-2, Factory Mutual Research, May 1971.

[73] Veldman, C. C., Kubota, T., and Zukoski, E. E., "An Experimental Investigation of the Heat Transfer from a Buoyant Gas Plume to a Horizontal Ceiling, Part I. Unobstructed Ceiling," Progress Report, Grant No. 5-9004, National Bureau of Standards, March–June 1975.

[74] Zukoski, E. E. and Kubota, T., "An Experimental Investigation of the Heat Transfer from a Buoyant Gas Plume to a Horizontal Ceiling, Part II. Effects of Ceiling Layer," Progress Report, Grant No. 5-9004, National Bureau of Standards, June–Sept. 1975.

[75] Moriya, T., Jin, T., and Shimade, H., "Study on Heat Transfer Coefficient in Hot Air and Emissivity of Flame in Burning Wooden House," Report No. 33, Fire Research Institute, Japan, 1971.

[76] Moriya, T., Shimada, H., and Morikawa, T., "Heat Transfer Coefficient from Fire in a Reinforced Concrete Building," Report No. 35, Fire Research Institute, Japan, 1972.

[77] Bromberg, K. and Quintiere, J., "Radiative Heat Transfer from Products of Combustion in Building Corridor Fires," NBSIR 74-596, National Bureau of Standards, Washington, D.C., Feb. 1975.

[78] Kashiwagi, T., "A Study of Flame Spread over a Porous Material under External Radiation Fluxes," Fifteenth Symposium (International) on Combustion, The Combustion Institute, 1974, pp. 255–265.

[79] Dayan, A. and Tien, C. L., Combustion Science and Technology, Vol. 9, 1974, pp. 41–47.

[80] de Ris, J. and Orloff, L., "The Role of Buoyancy Direction and Radiation in Turbulent Diffusion Flames on Surfaces," Fifteenth Symposium (International) on Combustion, The Combustion Institute, 1974, pp. 175–182.

[81] Simms, D. L., Combustion and Flame, Vol. 4, 1960, pp. 293–300.

[82] Kanury, A. M., "Ignition of Cellulosic Solids—A Review," FMRC Serial No. 19721-7, Factory Mutual Research, Norwood, Mass., Nov. 1971.

[83] Masliyah, J. H. and Steward, F. R., Combustion and Flame, Vol. 13, Dec. 1969, pp. 613–625.

# DISCUSSION

*C. T. Grimm*[1] (*oral discussion*)—Is the rate of burning significantly affected by the thermal storage capacity of the structure?

*James Quintiere* (*author's oral closure*)—Yes, it is. This is displayed by the model. I should point out that, in doing a theoretical analysis like this, we just tend to get trends and qualitative effects; they are semiquantitative; they give us the right order of magnitude. But we really need to do some experimental work to assess that property. There are a lot of parameters in this model that have been fixed, and there is uncertainty with

[1] Civil Engineering Department, University of Texas, Austin, Tex. 78798.

them. So yes, the model does display this effect, but whether in the correct amount, I don't know at the moment.

*H. Mitler*[2] (*oral discussion*)—I was curious as to whether you tried to compare your theoretical results to actual experimental data?

*James Quintiere* (*author's oral closure*)—If I were to do that I would have to consider the transient problem, and I have not gone to that point as yet with this model, but I know people who are pursuing that in other organizations. I've looked for qualitative trends from a host of experiments, and they tend to be somewhat in agreement with this model. One thing that this model does not provide in its present form is an indication of the enhancement of burning rate in a compartment over that with the same amount of fuel in the free burning state. So it's deficient there.

*I. Glassman*[3] (*oral discussion*)—I certainly think the model is useful for predicting the ventilation rate with fuel controlled fires, but I become very concerned when I hear you try to apply it to flashover problems. First of all, as I understand it, flashover involves a flame spreading process; there is a release of energy at one point that propagates (like a flame) in a given direction or in many directions. If so, it is difficult for me to see how this process could occur in your model when you uniformly take the temperature at the ceiling as roughly 600°C. Secondly, it seems that if you are going to talk about flashover, I feel very deeply about this problem, we are talking about limit (flammability) mixtures, and along the roof you will eventually get within a flammability limit. And when there is an ignition source, indeed the original flame or, as you said before, the hot products, get to a high enough temperature, they gasify to a limit mixture and are ignited at a point. Since you are within an enclosure and have stratified flow, then you should get rapid propagation which I think some call flashover, but I don't see how that can come out of your model.

*James Quintiere* (*author's oral closure*)—The gas phase combustion process would not be contained in my model. My model is primarily a thermal model. One can view, I feel, flashover or the development of flashover through several mechanisms. One mechanism is the thermal mechanism which is heat transferred to other items causing them to ignite or causing the fuel that is first ignited to burn more rapidly due to radiation feedback back to it or causing the fire to spread away from it more rapidly due to radiation enhancement. These mechanisms can be expressed by this kind of model and, in that sense, it's a fire growth process.

*I. Glassman* (*oral discussion*)—It's not a propagation of the flame?

*James Quintiere* (*author's oral closure*)—No, it's not the gas phase combustion problem that may be one of the mechanisms of flashover, and it comes down to this question of really how do you define flashover.

[2] Research associate, Harvard University, Cambridge, Mass. 02138.
[3] Professor, Department of Aerospace and Mechanical Science, Princeton University, Princeton, N. J. 08540.

*J. A. Rockett*[4] (*oral discussion*)—Relative to Professor Glassman's comment about the flame in the gas phase "along the roof," one of the more intractible problems at the present time is the physical extent of the combusting envelope and the degree of combustion achieved. Room burning behavior will be qualitatively different if the combusting plume impinges or does not impinge on the ceiling, and the effect will be exaggerated for a combustible ceiling. Further, flame height is known to be increased by an adjacent wall. Aside from the work of Steward (1964) there is little guidance on how to estimate turbulent flame heights. Work by Faeth, now in progress at Penn State, will give us some idea of the effect of both an adiabatic and an isothermal wall on plume characteristics. Theoretical work by Baum here at the National Bureau of Standards should improve our ability to deal with turbulent mixing in bouyant plumes. From this, a better technique for estimating flame heights and completeness of combustion could be obtained. Nevertheless, there is need for considerable work in this area before we will be able to predict the gas phase burning behavior and its effect on room flashover.

[4]Chief, Program for Physics and Dynamics, National Bureau of Standards, Washington, D.C. 20234.

*Takao Wakamatsu*[1]

# Calculation Methods for Predicting Smoke Movement in Building Fires and Designing Smoke Control Systems

**REFERENCE:** Wakamatsu, Takao, "**Calculation Methods for Predicting Smoke Movement in Building Fires and Designing Smoke Control Systems,**" *Fire Standards and Safety, ASTM STP 614,* A. F. Robertson, Ed., American Society for Testing and Materials, 1977, pp. 168–193.

**ABSTRACT:** This paper proposes a basic direction for building systems designs based on scientific methods of quantitative evaluation in order to ensure human safety in building fires. Herein are described the position of the smoke control system as a subsystem of the whole system and the methods of designing the subsystem. Design of the subsystem is intended to produce a rational smoke control system best suited to each individual building based on computer simulation predictions of smoke flow or pressure distribution in the building. An outline of calculations on flowrate and pressure, concentration of smoke and gases, and thermal conditions is given. Calculations are divided into three steps according to the required level of accuracy, and these steps may be used appropriately, depending on specific goals. As an example of their application, the author demonstrates the design of a simplified smoke control system method, showing that it can provide simple, quantitative evaluation of the effectiveness of various smoke control systems under different conditions.

**KEY WORDS:** fires, fire safety, smoke, ventilation

It is an undeniable fact that fire disasters involving many fatalities have occurred frequently in Japan during recent years. Certainly, one can obtain many useful suggestions for safety measures from such individual fires, but such experience has not yet accumulated to the degree that it can be appropriately reflected in safety measures and techniques. Furthermore, there has been no methodology which reflects such experience in the effective introduction of safety technology and the development of fire safety systems. Accordingly, the development of theory and methods

[1] Chief of Fire Research Section, Building Research Institute, Tokyo, Japan.

for the quantitative evaluation of safety systems and the optimum design and operation of these systems is a vital task which must be performed in order to raise the level of fire prevention technology.

From such a standpoint, we have been actively concerned, for some years, with carrying on a variety of research programs to gain a scientific understanding of smoke production and movement mechanisms and establishing engineering design systems that can be rationally applied to building fire safety design. This paper outlines our research concept on such matters and introduces our general concept of human safety and smoke control systems in building fires, calculation methods for predicting smoke movement and designing smoke control systems, and an example of smoke control design.

## Systems for Human Safety in Building Fires

For the sake of convenience, we define fire prevention safety design as "design to derive the optimum system needed for each building to ensure the safety of its occupants from fires, based on engineering principles, that is, on quantitative evaluation." In evaluating and determining a safety system for a building through fire prevention safety design, freedom should be allowed in the selection of the methods and structure of the safety system, providing the required or target level of safety can be secured. In other words, the objective of the fire preventive safety design is to design a safety system which satisfies daily functional convenience and economic aspects and also guarantees the necessary safety without being limited to fixed methods and structures.

A conceptual diagram of a safety system applying such design is shown in Fig. 1. As is clear from this diagram, the safety system is grouped roughly into a train of subsystems to secure the safety of human lives (including the safety of property) by preventing the outbreak and spread of fire, and another train of subsystems to secure the safety of human lives after the outbreak of fire

$$R_T = 1 - [(1 - r_1)(1 - r_2)(1 - r_3)] \times [(1 - r_4)(1 - r_5)]$$

[probability that fire will break out and spread] ×

[probability that life will be in danger]

These subsystems themselves include subsystems for precautions against fire outbreak, limitation of fire growth, etc. The total safety system is composed of these subsystems in parallel form. Thus, if $R_T$ is the assessment of the overall reliability or effectiveness of this safety system, it can be expressed as follows. This equation shows that if any one $(r_i)$ of the five subsystems $(r_1 \sim r_5)$ is perfect $(r_i = 1)$, this safety system becomes

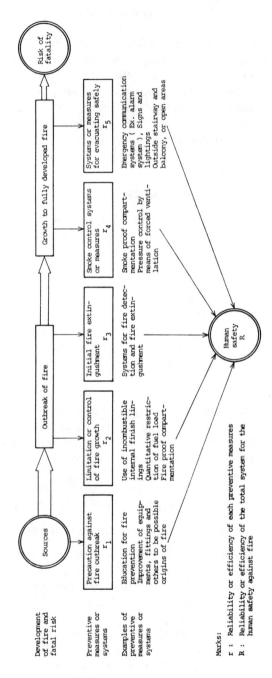

FIG. 1—*Life safety system against building fires.*

perfect ($R_T = 1$), even without the presence of other subsystems ($r_j$). Realizing, however, that such a perfect subsystem may be almost impossible, it may be necessary to devise the constitution of the subsystems (apportionment of $r_1 \sim r_5$) best suited to individual buildings so that the design will enhance the reliability $R_T$ of the total system, not sacrificing to apportionment the conveniences and economics of the buildings.

In some cases, partial quantitative assessment of the reliability (effectiveness) of each subsystem may be possible by statistical and probabilistic means and analyses of fire phenomena, but, in general, a number of difficulties will remain. Nevertheless, it seems very important to continue developing fire hazard research, keeping this objective in mind.

### Smoke Control System and Its Design

The smoke control system, as shown in Fig. 1, is ranked as one of the subsystems of the total safety system. The smoke control system is also made up of such subsystems as spatial construction of the building, ventilation equipment, and their starting and operating devices. The assessment of this system ($r_4$) is determined by the probability ($r_4'$) that equipments in the system will start and operate as expected (reliability of equipments) and by the probability ($r_4''$) that this will control the movement of smoke as well as designed (effectiveness of the system). Generally, ($r_4$) can be assessed as ($r_4 = r_4' \times r_4''$). The effectiveness ($r_4''$) of the system can also be defined as a percentage to show to what extent those conditions that were accounted for at the design stage have covered the probable fire conditions at the time of a real fire. In this sense, the method setting hypothetical conditions for use in its design is of great significance.

The movement of smoke in building fires essentially follows laws of physics, and, therefore, its physical process should be determined by ambient physical conditions. Thus, if the physical laws and conditions for the movement of smoke are clarified, it is eventually possible to predict the movement.

Regarding the former, as will be described later, it seems right to say that a calculation method for predicting the movement of smoke based on physical laws may now be in the final stage for practical application.[2-4] Also, the results of several field experiments confirm that the calculation method has enough reliability for practical use. However, as to the latter, it is extremely difficult to predict in advance the various conditions such as spatial construction, openings or flowpaths, external weather, and so

[2] Wakamatsu, Takao, "Experiments on Smoke Movement in Building," Occasional Report of Japanese Association of Fire Science and Engineering, No. 1, 1974.

[3] Wakamatsu, Takao, "Smoke Movement in Building Fires," Occasional Report of Japanese Association of Fire Science and Engineering, No. 1, 1974.

[4] Wakamatsu, Takao, "Unsteady-state Calculation of Smoke Movement in an Actually Fired Building," Proceedings, The Control of Smoke Movement in Building Fires, CIB W14 Symposium, International Conference of Building Officials, 1975.

on. That is, the difficulty in predicting these conditions make predicting smoke movement more difficult.

There are various means for controlling smoke movement. "Smoke control design" can be defined as a way to find a means or system of controlling the smoke which is well suited to each individual building and then to derive the optimum quantitative conditions required for those means. Smoke control design, up to the present, is neither widespread in its concept nor systematized as an engineering system. From now on, it may be necessary to arrange and establish a theoretical system for design with the following three items as its skeleton.

1. Data that provide hypothetical conditions for design in calculation.
2. Calculation methods that allow quantitative evaluation of the smoke control effect.
3. Methods of introducing the calculation results into the control systems.

Item 3 is necessary to assess the results obtained from Items 1 and 2 in light of the reliability of Items 1 and 2. In the case of Item 3, it will be necessary to arrange methods and standards for assessing reliability and establishing safety factors. Here, Items 1 and 2 are only used for detailed description because Item 3 is a future rather than a pressing problem compared with the other two.

**Hypothetical Conditions in Smoke Control Design**

It is almost impossible to predict when, where, and under what conditions a fire will occur. However, it may be possible to limit the scope of the probable conditions to some extent by studying the probabilities derived through analyzing data on statistics of fires and meteorology, actual conditions of various buildings, and other factors concerned. As for the conditions which have a great effect on smoke movement in buildings, that is,

1. Times of fire outbreaks.
2. Meteorological conditions (direction and velocity of wind, air temperature, wind pressure coefficient).
3. Location of fire outbreaks (floor and compartment of origin).
4. Flow resistances of doors, windows, ducts, and shafts (flow coefficients, ratio of broken windows and opened doors, etc.).
5. Temperature distribution in buildings (fire compartment, general compartments, staircases, ducts, and shafts).

it may be adequate at the present time to establish a standard for each of these conditions for input data so that the designed system is always on the safe side.

For the sake of convenience in designing the control systems in Japan, we offered the following tentative proposals for the five conditions just described. However, many of these are not supported statistically or experimentally, and there remains much future study to be done, especially the conditions of open doors and windows at the time of fire outbreaks.

### Times of Fire Outbreaks

Wintertime (buildings being heated) and summertime (buildings being air conditioned), during which a great temperature difference exists between the external temperature and the air within buildings should be chosen as a design condition.

### Meteorological Conditions

They should be determined as follows on the basis of the meteorological statistics in each particular region during winter and summer.

*Air Temperature*

Winter     Minimum average monthly values of the daily lowest temperatures (example: $-1.5\,°C$, January, Tokyo)

Summer     Maximum average monthly values of the daily highest temperatures (example: $30.7\,°C$, August, Tokyo)

*Wind Velocity*—Wind velocity $V_0$, up to that which encompasses 95 percent of accumulated wind velocity frequency observed at the observatory in each area during winter and summer, is set as a standard, and wind velocity $V$, which corresponds to the arbitrary height $h$ of a building, is determined as follows according to the site of each building.

For high-rise buildings built on broad sites

$$V = k \times V_0(h \leq h_0)$$
$$V = k \times V_0(h/h_0)^{1/k}(h > h_0)$$

For buildings built within urban areas

$$V = k \times V_0(h \leq h_0),$$
$$V = k \times V_0(h/h_0)^{1/k}(h > h_0)$$

where

$k$ = revising factor for wind velocity and

$h_0$ = height from the ground to the place where the anemometer of each observatory is located.

*Wind Pressure Coefficient and Wind Direction*—The standard wind direction is taken so that it is normal to the principal openings or (windows) and hence causes positive wind pressure on them. The wind pressure coefficient for each opening facing the external air is decided on the basis of the standard wind direction.

### Location of Fire Outbreaks

As for the determination of floors where a fire may occur, consideration must be given to the natural ventilation effect on smoke movement upward or downward through shafts used particularly during heating or air-conditioning periods. Taking this into consideration, studies should be made on lower floors (for example, basement, ground floor, second floor) for winter and on both upper floors (for example, top floor) and lower floors for summer. In many cases, there will be a floor which will bring the most dangerous smoke spread in the event of fire. In addition to such standard cases, more work is required to study the situations posed by changes in the floor of origin, as in a building with special features in its spatial construction.

With regard to the compartment of origin, attention should be centered on those which satisfy the following requirements on the specific floor.

1. A compartment which is densely occupied, being used for general purposes.

2. The compartment considered to have the highest probability of fire outbreak.

3. The compartment estimated to present the greatest difficulty in smoke control should a fire occur in it. When there are more than two compartments which satisfy these conditions, and it is predictable that a certain smoke control measure may produce considerably different results in each, a study is made on each of them.

### Flow Resistance of Openings (Windows and Doors) and Ducts and Shafts

Concerning an opening (or a flowpath) detailed as to cross-sectional conditions, the value of its resistance can be easily obtained from such data as the ventilation design standards at room temperature. Although it is very difficult to estimate the opening conditions of windows and doors in fires, a daring assumption is proposed for this problem, as shown in Table 1. In this table, for example, the figure "zero" or "one" signifies that the door (or window) is normally closed, or fully opened (or broken), respectively.

### Temperature Distribution Buildings

This is a steady-state calculation assuming that there is no temperature change with time in a part of a building. Table 2 shows the hypothetical temperature fixed for each part of the building. The temperature of a

TABLE 1—*Opening ratio for design (ratio of effective flow area to fully opened area).*

| Opening | Position and Specification of Opening | Fire Room | Escape Route | General Part |
|---|---|---|---|---|
| Window | common glass | 0.5 | 0.2 | 0 |
| | wired glass | 0.2 | 0.2 | 0 |
| | heatproof glass | 0 | 0.2 | 0 |
| Door | with device for self-closing | | | |
| | steel | 0.1 | 0.5 | 0 |
| | wooden | 1 | 0.5 | 0 |
| | others | | | |
| | steel | 0.5 | 0.8 | 0.5 |
| | wooden | 1 | 0.8 | 0.5 |
| Steel shutter | with heat detector | 0.5 | 1 | 1 |
| | with smoke detector | 0.2 | 1 | 1 |
| | manual operation | 0.8 | 1 | 1 |

TABLE 2—*Temperature assumed for design (°C).*

| Compartment | Winter | Summer |
|---|---|---|
| Fire room | 900 | 900 |
| General room | 20 | 25 |
| Staircase | 15 | 27 |
| Duct, shaft[a] | 10 ~ 30 | 10 ~ 30 |

[a] Adopt usual temperature according to the use of the duct of shaft.

fire room or compartment shown in this table is determined to be a fully developed fire which, in general, makes the designed system on the safe side.

## Calculation Method of the Movement of Smoke and Smoke Control

The calculation method for predicting the fluid phenomena of smoke and air ranges from microscopic to macroscopic levels. Depending on whether the fluid field that is in question deals with the whole building or an extremely localized part of it, the level of the calculation method for use in its analysis is eventually changed. For instance, to solve the fluid characteristics in full detail for a fire compartment at the initial stage of a fire, numerical calculation of basic fluid dynamics equations by means of the finite difference method is essential, and consideration has to be given to compressibility as a computation factor.

The method adopted here, so to speak, is the most macroscopic one, in which a whole building is regarded as a fluid field for the calculation.

That is to say, this method intends to solve the movement of smoke and air in a building as a whole by computing the quantity of smoke and air flowing through each opening and flowpath, on the assumption that the temperature is uniform in each compartment except the vertical shafts, and without giving special thought to the localized flows in the building, for example, inside a certain room.

When calculation is made of smoke movement and its control, the pressure and flowrate in each part of a building must be computed, and, as the case demands, it also becomes necessary to carry out calculations for the concentration of smoke and gases, and heat in order to obtain the temperatures in each part of the building.

## Calculation of Pressure and Flow Rate

Under the same conditions of temperature and pressure, there is no remarkable difference in fluid properties between smoke and air. Consequently, it is convenient to regard smoke in the same light as air and calculate only the flow of air, and, as the need arises, to consider the concentrations of smoke or gases included in the air. Accordingly, in this section, calculation of pressure and flowrate is made only on air.

When a pressure difference $\Delta P$ across an opening or a flowpath is given, the mass flowrate $Q$ can be obtained generally from the following equation

$$Q = \alpha A \sqrt{2g\gamma|\Delta p|} \tag{1}$$

where

$\alpha$ = flow coefficient, $\alpha = 1/\sqrt{\xi}$,
$\xi = 1 + \zeta + \lambda(L/D)$,
$\zeta$ = pressure loss coefficient of flowpath (or opening),
$\lambda$ = frictional resistance coefficient of flowpath,
$L$ = length of flowpath,
$D$ = equivalent diameter of flowpath,
$A$ = sectional area of flowpath (or opening),
$g$ = gravitational acceleration, and
$\gamma$ = density of air.

Since there may be cases, such as those shown in Fig. 2, where the pressure difference at the opening in question is not uniform because of the temperature difference between the two spaces $i$ and $j$ on each side of the opening, Eq 1 is more generalized and is expressed in the following equation

$$Q = K\alpha B(h_2 - h_1)\sqrt{2g\gamma|\Delta P_a|} \tag{2}$$

where for $\gamma_i = \gamma_j$

$$K = 1, h_2 - h_1 = H_h - H_l, \Delta P_a = \Delta P_{ij}$$

$$\gamma = \gamma_i (\Delta P_{ij} > 0), \text{ or } \gamma = \gamma_j (\Delta P_{ij} < 0)$$

and where for $\gamma_i \neq \gamma_j$

$$K = \frac{2\sqrt{2}}{3} \sqrt{1 + \frac{n}{(1 + n)(1 + \sqrt{n})^2}}, \quad \text{where } n = h_1/h_2$$

$$\Delta P_a = \frac{h_1 + h_2}{2} \Delta\gamma_{ji}, \quad \text{where } \Delta\gamma_{ji} = \gamma_j - \gamma_i$$

The values of $h_1$, $h_2$, and $\gamma$ for $\gamma_i \neq \gamma_j$ in Eq 2 can be obtained from Table 3 and Fig. 2, depending on the height of the neutral plane $Y_N$ and on

TABLE 3—Relationship between $h_1$, $h_2$, $\gamma$ and position of neutral plane.

| Position of Neutral Plane | Flow Part of an Opening | $h_1$ | $h_2$ | $\gamma$ $\gamma_o > \gamma_i$ | $\gamma_o < \gamma_i$ |
|---|---|---|---|---|---|
| $Y_N \geqslant H_h$ | whole area | $Y_N - H_h$ | $Y_N - H_l$ | $\gamma_o$ | $\gamma_i$ |
| $H_l < Y_N < H_h$ | upper part | 0 | $H_h - Y_N$ | $\gamma_i$ | $\gamma_o$ |
| | lower part | 0 | $Y_N - H_l$ | $\gamma_o$ | $\gamma_i$ |
| $H_l \leqslant Y_N$ | whole area | $H_l - Y_N$ | $H_h - Y_N$ | $\gamma_i$ | $\gamma_o$ |

whether the value of $\Delta\gamma_{ji}$ is positive or negative. In addition, according to Fig. 2, $Y_N$ is given by the following equation

$$Y_N = \Delta P_{ji}/\Delta\gamma_{ji}$$

When Room $i$ is considered as a subject, let $Q_{ji}$ be the mass rate of air

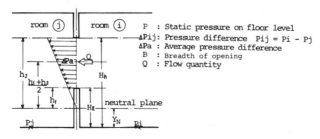

FIG. 2—Mathematical model of flow quantity, pressure, and opening conditions.

flow via an arbitrary Opening $o$ from an arbitrary adjacent Room $j$ to the Room $i$, and let $Q_{ij}$ be the mass rate of air flow from $i$ to $j$. Then, absolute values of $Q_{ij}$ and $Q_{ji}$ are obtained from Eq 2. Let $Q_{ij} \leqq 0$ and $Q_{ji} \geqq 0$; then the apparent mass rate of air flow through Opening $o$ into Room $i$ is expressed by the following equation

$$Q_{io} = (Q_{ij} + Q_{ji})_o \tag{3}$$

Here, if $\gamma_j = \gamma_i$, or $Y_N \leqq H_l$ or $Y_N \geqq H_h$ for $\gamma_j \neq \gamma_i$, either $Q_{ij}$ or $Q_{ji}$ eventually becomes zero.

This is the gist of a calculation method for obtaining the mass rate of air flow via an opening where the pressures and temperatures (or specific density $\dot{\gamma}$) in two rooms located on both sides of the opening are given. However, in practice, the pressure in each room of a building is generally unknown, so, in order to obtain the pressure and, simultaneously, the flow quantity, the mass balance in each room must be considered.

Now, let us consider an arbitrary Room $i$ connected with Rooms $m$ in number, via flowpaths $m'$ in number for any of the rooms. Then, the mass balance $\Sigma Q_i$ of Room $i$ is defined from Eq 3 as follows

$$\Sigma Q_i = \sum_{j=1}^{m} \sum_{o=1}^{m'} (Q_{ij} + Q_{ji})_{o,j} \tag{4}$$

$Q_{ji}$ and $Q_{ij}$ also include the amount of forced ventilation. On the other hand, let $P_i$ and $P_j$ be the static pressure at the standard level (for example, floor level) in Rooms $i$ and $j$, respectively, then the flowrates $Q_{ij}$, $Q_{ji}$ are described by the equations

$$Q_{ij} = f_{ij}(P_i, P_j) \quad \text{and} \quad Q_{ji} = f_{ji}(P_i, P_j)$$

Accordingly, Eq 4 is written as

$$\Sigma Q_i = f_i(P_i, P_1, P_2, \ldots, P_m)$$

In a case when there are compartments $n$ in number in a building, the mass balance equation for incompressible fluid for each compartment is expressed by

$$\Sigma Q_i = f_i = 0$$

or

$$f_1(P_1, P_2, \ldots, P_n) = 0$$
$$f_i(P_1, P_2, \ldots, P_i, \ldots, P_n) = 0 \tag{5}$$
$$f_n(P_1, P_2, \ldots, P_n) = 0$$

That is to say, for $P_1$, ..., $P_n$ is obtained by solving simultaneously $n$ number of mass balance equations with unknown quantities $n$ in number, that is, pressures $P_1$, ..., $P_n$, in each room. The pressures to be included in an equation $f_i = 0$ are, in practice, only $P_i$ and the pressures in those rooms connected direclty to Room $i$ via some flowpaths or openings.

Since each of the simultaneous Eqs 5 becomes a nonlinear equation as represented in Eq 2, it is generally difficult to solve Eq 5 in an analytical way. An example of its solution is shown next.

First of all, from the first equation in Eqs 5, $P_1$ is given as a function of $P_2$, ..., $P_n$, that is

$$P_1 = f_1*(P_2, P_3, \ldots, \overset{\rightharpoonup}{P_n})$$

Substituting this into $f_2 = 0$, ..., $f_n = 0$ in Eqs 5, simultaneous equations $(n - 1)$ in number are obtained, from which $P_1$ is eliminated, that is

$$f_2[f_1*(P_2, P_3, \ldots, P_n), P_2, P_3, \ldots, P_n] = 0$$
$$f_n[f_1*(P_2, P_3, \ldots, P_n), P_2, P_3, \ldots, P_n] = 0 \tag{6}$$

Successive elimination of $P_2$, ..., $P_{n-1}$ by a similar procedure turns Eqs 5 into the following equation consisting of only $P_n$

$$f_n*[f_1* \cdot f_2* \cdot \ldots \cdot f_{n-1}*(P_n), f_2* \cdot f_3* \cdot \ldots \cdot f_{n-1}*(P_n), \ldots,$$
$$f_{n-2}* \cdot f_{n-1}*(P_n), f_{n-1}*(P_n), P_n] = 0 \tag{7}$$

where, for example,

$$f_1* \cdot f_2* \cdot f_3*(P) = f_1*(f_2*(f_3*(P)))$$

A flow chart demonstrating this procedure is shown in Fig. 3.

A general method to obtain the pressure in each compartment and the flow quantity in each flowpath in a building was described previously. However, in practical applications, it is possible to calculate this using a simpler procedure, depending upon the spatial construction of a given building.

## Calculation of Smoke and Gas Concentrations in a Compartment

Considering the balance of the concentrations at an arbitrary time $t = k\Delta t(k = 0, 1, 2, \ldots)$ during a time interval $\Delta t$ with regard to an arbitrary Compartment $i$, the following balance equation is obtained

$$[\sum_j (Q_{ji} \times C_j) - \sum_j (Q_{ij} \times C_i)] \Delta t = (V\gamma\Delta C)_i \tag{8}$$

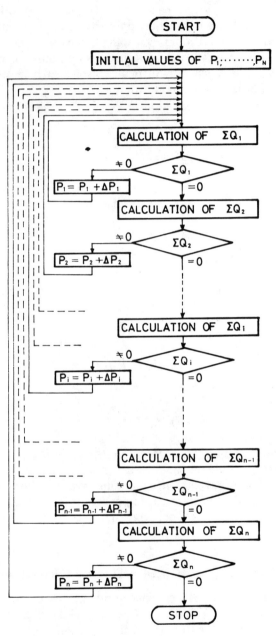

FIG. 3—*General procedure for calculation mass balance equations in a building.*

where

$V$ = air volume,
$\gamma$ = specific density of air, and
$\Delta C_i$ = a change of concentration in the time interval $\Delta t$ in the compartment, that is, $\Delta C_i = C_i(k + 1) - C_i(k)$

From the mass balance in the compartment, expressing $\Sigma Q_i \equiv \Sigma Q_{ji} = \Sigma Q_{ij}$, and $Q_{ij}$, $Q_{ji}$, etc., at $t = k\Delta t$, respectively, by $Q_{ij}(k)$, $Q_{ji}(k)$, etc., the following equation is obtained from Eq 8

$$C_i(k + 1) = \frac{\Sigma [Q_{ji}(k) \times C_j(k)]_j}{\Sigma Q_i(k)} + \left\{ C_i(k) - \frac{\Sigma [Q_{ji}(k) C_j(k)]_j}{\Sigma Q_i(k)} \right\} \times$$

$$\exp[- \frac{\Sigma Q_i(k)}{V_i \gamma_i(k)} \Delta t]$$

(9)

The value $C_i(k)$ of the smoke or a gas concentration in a compartment can be expressed in terms relative to the concentration of $C_F$ in a fire compartment, providing that $C_F$ can be assumed to be constant (for example $C_F = 1$) during the fire.

## Heat Calculation

Regarding heat calculation to obtain the temperature in each compartment, heat balance during a time interval $\Delta t$ at $t = k\Delta t$ in Compartment $i$ is considered in a manner similar to that described previously for the calculation of concentration. The heat balance equation is described by the equation

$$\left\{ C_p[\Sigma_j(Q_{ji} \times T_j) - \Sigma_j (Q_{ij} \times T_i)] - \Sigma_w[A_{iw} \times h_{iw}(T_i - T_{iw})] \right\} \Delta t = (V\gamma C_p \Delta T)_i$$

(10)

where

$C_P$ = specific heat of air at constant pressure,
$T_i$, $T_j$ = temperature in Compartments $i$ and $j$, respectively, where $j$ is a compartment connected with $i$,
$\Delta T$ = a change of temperature in a time interval $\Delta t$,
$A_{iw}$ = surface area of arbitrary wall (or ceiling or floor) $w$ in Compartment $i$,
$T_{iw}$ = surface temperature of the wall, and
$h_{iw}$ = heat transfer coefficient from Air $i$ to the Wall $w$.

Expressing the values concerned with a time $t = k\Delta t$ in the method used for Eq 9, the following equation is obtained from Eq 10

$$T_i(k + 1) = T_i(k) \{ 1 - \frac{\Delta t}{[V \times \gamma(k)]_i} \sum_j Q_{ij}(k) \} + \frac{\Delta t}{[V \times \gamma(k)]_i} \times$$

$$\{ \sum_j [Q_{ji}(k) \times T_j(k)] - \frac{1}{C_P} \sum_w A_{iw} \times h_{iw}(k) \times [T_i(k) - T_{iw}(k)] \} \tag{11}$$

There are several methods to calculate the surface temperature $T_{iw}$ of a surrounding wall (or ceiling, floor, etc.) $w$. And, in cases where the wall is assumed to be a semi-infinite solid, the following equation is obtained, assuming $T_{iw}(0) = T_i(0)$

$$T_{iw}(k) = T_i(1) - [T_i(1) - T_i(0)] \exp[H_{iw}^2(0) \times \varkappa_{iw}\Delta t] \times \text{erfc}[H_{iw}(0) \times$$

$$\sqrt{\varkappa_{iw} \times \Delta t}] + \ldots + [T_i(k) - T_i(k - 1)] \{ 1 - \exp[H_{iw}^2(k - 1) \times \tag{12}$$

$$\varkappa_{iw}\Delta t] \times \text{erfc}[H_{iw}(k - 1)\sqrt{\varkappa_{iw} \times \Delta t} \}$$

where

$H_{iw} = h_{iw}/\lambda_{iw}$,
$\lambda_{iw}$ = thermal conductivity of Wall $w$, and
$\varkappa_{iw}$ = thermal diffusivity of Wall $w$.

## Kinds of Calculation Methods and Their Application

The calculation for predicting smoke movement or smoke control design can be made on the basis of the three kinds of calculations just described, which are suitably combined to meet specific requirements. For example, when a calculation is made for smoke control design, the calculation of the smoke concentration is not necessarily required, but, in many cases, it is sufficient for the design to calculate only pressure conditions required for smoke control. In other words, such calculations each differ in the accuracy and output data required or appropriateness of labor and time to be spent. Therefore, it seems convenient to categorize the kinds and application of calculation methods, that is, unsteady-state, precise steady-state, and simplified steady-state calculation methods.

The aim of the unsteady-state calculation method is to clarify the changes with time of the flowrates, concentrations, and temperatures in each part of a building in response to the changes in conditions which vary from time to time after the fire breaks out. To apply this calculation method, all of the three formulas just mentioned have to be solved simultaneously. This calculation is advantageous in that is can predict or clarify not only the movement of smoke but also the thermal conditions in a building with more accuracy than the others following, while, on the other hand, it is disadvantageous in that it generally demands much more time than the others. As applied examples of this calculation method, several analyses have been carried out for actual fires occurring in Japan (for instance, the Sennichi Department Store, a hospital, and field fire experiments). Some of these have already been reported in publications.

The precise steady-state calculation method is a method of solving the mass balance equations for all spaces in a building, assuming that the conditions of the openings, external weather, and temperatures in the building do not change with time, in other words, that the movement of air and smoke is regarded as steady. Therefore, this method enables one to accurately predict the movement when various conditions which may exert a serious influence upon the movement can be regarded as approximately steady, for instance, as is seen with a fully-developed fire in a fire compartment. In practical application, this seems fairly effective when highly accurate evaluation is required in the smoke control design on the assumption of a steady-state flow (assumption on the safe side). It is, of course, possible to calculate the concentration of smoke or gases in individual compartments on the basis of the results on flowrates obtained through this method. Further, this can be said to be the basic calculation method on which the aforementioned and the hereafter-described methods are based.

The simplified steady-state method is defined here as a method for solving the mass balance equations only, for spaces such as each space on a floor of origin and vertical shafts, where the major flows are formed. In this case, flow resistances of omitted spaces are synthesized together. One variety of this method is to immediately obtain the ventilation conditions required for the intended smoke control by introducing an equation on smoke stopping conditions. This method will be described in more detail in the next section.

Since it is possible to conduct these calculations with exceedingly short computational time, this method may be very useful in cases where a great number of calculations on different conditions are required for a certain building. It will be more practical, in designing the smoke control systems, to evaluate the efficiencies of many systems posed for the control under various conditions at an initial stage of the design, and then to recheck some systems (estimated to be realized through the calculations) by means of any of the more precise methods mentioned previously at a final stage.

### Case Study on Smoke Control Design

Many efforts have been made to control the movement of smoke in fires by means of natural or mechanical ventilation, and many proposals have been made on the control methods for smoke. However, because smoke movement is greatly influenced by external conditions (air temperature, wind), conditions of a building (spatial construction, opening conditions, temperature distribution in each space, etc.), it is difficult to ensure uniformly the safety of all individual buildings by means of a certain method of system. Therefore, by comparing the methods or systems posed for a building with a quantitative evaluation of their efficiencies under different conditions, the most suitable one may be chosen from them and the requirements for the method or system derived, for example, as to sizes of ventilation openings, amount of forced ventilation, and so on. For these reasons, some calculation methods are necessary for designing the smoke control systems.

Mainly to explain the calculation procedure use of the simplified method in more detail, a case study is introduced here, taking the following building model as an example.

### Model for Calculation

A mathematical model for the simplified method is shown in Fig. 4, focussing on the floor of origin, on which fire room ($F$), corridor ($C$),

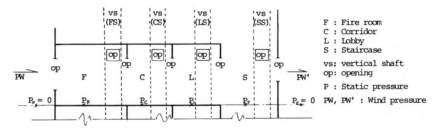

FIG. 4—*A model for simplified calculation* (*sectional elevation of fire floor*).

lobby ($L$), and staircase ($S$) are arranged in this order. Each of the compartments has two openings, one facing the open air and the other connecting the compartment to a vertical shaft (*FS, CS, LS, SS*). The vertical shaft here represents an air-conditioning duct system, elevator shaft, smoke tower, etc.

### Formation of Smoke Control

It is idealistic to confine the spread of smoke only to the compartment of origin and ensure steady safety of the occupants in the rest of the building from the smoke and gases in the event of fire.

The location of measures to cut off the smoke spread should be decided in relation to such factors as emergency evacuation systems. In this model, the openings or doors lying between *F-C, C-L,* and *L-S* can be considered as the prearranged position for stopping smoke on the fire floor, and all of these positions are taken into consideration in the calculation. On the other hand, as the prearranged position for forced ventilation, the compartments *F, C, L, S,* and the combinations of them can be considered as shown in Table 4, even if most of them are not practical. In the design of smoke control, it may be sufficient to choose some practicable combinations from among these four.

TABLE 4—*Combination of forced ventilations.*

| Number of Position for Forced Ventilation | Position for Forced Ventilation | | | |
|---|---|---|---|---|
| | Fire Room | Corridor | Lobby | Staircase |
| 1 | − | | | |
| | | + − | | |
| | | | + − | |
| | | | | + |
| 2 | − | + − | | |
| | − | | + − | |
| | − | | | + |
| | | + − | + − | |
| | | + − | | + |
| | | | + − | + |
| 3 | − | + − | + − | |
| | − | + − | | + |
| | − | | + − | + |
| | | + − | + − | + |
| 4 | − | + − | + − | + |

NOTE— + means air supply, − means smoke exhaust; + − means + or −.

## Procedure of Calculation

While there are several simplified-method smoke control designs, a method which enables one to obtain very easily the required amount of forced ventilation will be introduced here as an example. In this example, only such principal spaces as the compartments (*F, C, L, S*) and the vertical shafts (*FS, CS, LS, SS*) are used for calculation. Vertical shafts and spaces generally have a significant effect on controlling the smoke movement. It is assumed in this model that each of the shafts is connected with a synthesized resistance of flow to the external air. The pressures on a standard level in these principal spaces are represented by $P_F$, $P_C$, $P_L$, $P_S$, $P_{FS}$, $P_{CS}$, $P_{LS}$, and $P_{SS}$. Table 5 shows the relation between each mass balance equation and the pressures concerned with the equation,

TABLE 5—*Relationship between mass balance equation and pressure as variable.*

| $\Sigma Q$ \ $P$ | $P_F$ | $P_C$ | $P_L$ | $P_S$ | $P_{FS}$ | $P_{CS}$ | $P_{LS}$ | $P_{SS}$ |
|---|---|---|---|---|---|---|---|---|
| $\Sigma Q_F$ | O | O | | | O | | | |
| $\Sigma Q_C$ | O | O | O | | | O | | |
| $\Sigma Q_L$ | | O | O | O | | | O | |
| $\Sigma Q_S$ | | | O | O | | | | O |
| $\Sigma Q_{FS}$ | O | | | | O | | | |
| $\Sigma Q_{CS}$ | | O | | | | O | | |
| $\Sigma Q_{LS}$ | | | O | | | | O | |
| $\Sigma Q_{SS}$ | | | | O | | | | O |

where the pressures are regarded as variable in the equation. Since forced ventilation in each shaft is not taken into account, the pressure in each shaft is expressed, respectively, from the mass balance equation for each shaft, as a function of $P_F$, $P_C$, $P_L$, or $P_S$, given below

$$P_{FS} = f^*(P_F), \quad P_{CS} = f^*(P_C), \quad P_{LS} = f^*(P_L), \quad P_{SS} = f^*(P_S) \quad \dots \quad (13)$$

Consequently, in the composition of the calculation procedure, only $P_F$, $P_C$, $P_L$, and $P_S$ are taken into account as the pressure to be solved. The composition for each case differs slightly by the position of the smoke stopping and the forced ventilation. Therefore, the following shows an example to explain in more detail the procedure and composition for a case intending to obtain the required amount $\bar{Q}_F$ of forced ventilation in the fire Room $F$ for stopping smoke at each position. (In this case, $\bar{Q}_F$ seems naturally to be the amount of smoke to be exhausted.)

### The Case of Smoke Stopping at F-C

The simultaneous equations for this case consist of an equation for the required condition to stop the smoke and the mass balance equations for each compartment, and they are given in the following functional form

$$P_c = f(P_F) \quad \dots \text{ for smoke stopping} \tag{14}$$

$$\Sigma Q_F = f_F(P_F, P_C, \bar{Q}_F) = 0 \tag{15}$$

$$\Sigma Q_C = f_C(P_F, P_C, P_L) = 0 \tag{16a}$$

$$\Sigma Q_L = f_L(P_C, P_L, P_S) = 0 \tag{17a}$$

$$\Sigma Q_S = f_S(P_L, P_S) = 0 \tag{18}$$

By substituting Eq 14 in Eqs 16a and 17a, Eqs 16b and 17b are obtained

$$\Sigma Q_C = f(P_F, f(P_F), P_L) = 0 \tag{16b}$$

$$\Sigma Q_L = f(f(P_F), P_L, P_S) = 0 \tag{17b}$$

Here, expressing $P_F$ and $P_S$ as functions of $P_L$, from Eqs 16b and 17b, such as

$$P_F = f_C^*(P_L) \quad \text{and} \quad P_S = f_S^*(P_L)$$

and then, by substituting them in Eq 17b, Eq 17c is obtained as

$$f_L\{(f(f_C^*(P_L)), P_L, f_S^*(P_L)\} = 0 \tag{17c}$$

Thus, all of the unknown pressures $P_F$, $P_C$, $P_L$, and $P_S$ can be obtained from these formulas, and then, $\bar{Q}_F$—the amount of forced ventilation at the fire room required to stop the smoke at the opening or door $F$-$C$—is given by substituting $P_F$ and $P_C$ as obtained in Eq 15.

*The Case of Smoke Stopping at* C-L

$$P_L = f(P_C) \ldots \text{ for smoke stopping} \tag{19}$$

$$\Sigma Q_F = f_F(P_F, P_C, \bar{Q}_F) = 0 \tag{20}$$

$$\Sigma Q_C = f_C(P_F, P_C, P_L) = 0 \tag{21}$$

$$\Sigma Q_L = f_L(P_C, P_L, P_S) = 0 \tag{22a}$$

$$\Sigma Q_S = f_S(P_L, P_S) = 0 \tag{23a}$$

By substituting Eq 19 in Eqs 22a and 23a, Eqs 22b and 23b are given as functions of only $P_C$ and $P_L$

$$\Sigma Q_L = f_L(P_C, f(P_C), P_S) = 0 \tag{22b}$$

$$\Sigma Q_S = f_S(f(P_C), P_S) = 0 \tag{23b}$$

Here, expressing $P_C$ as a function of $P_S$ from Eq 22b, such as $P_C = f_L^*(P_S)$ and then, by substituting this into Eq 23b, Eq 23c is obtained as

$$f_S\{f(f_L^*(P_S)), P_S\} = 0 \tag{23c}$$

$P_F$, $P_C$, $P_L$, and $P_S$ can all be obtained from these formulas, and then, $\bar{Q}_F$—the required amount for smoke stopping at $C$-$L$—is given by solving Eq 20 with $P_F$ which is to be obtained by substituting $P_C$ and $P_L$ into Eq 21.

*The Case of Smoke Stopping at* L-S

$$P_S = f(P_L) \ldots \text{ for smoke stopping} \tag{24}$$

$$\Sigma Q_F = f_F(P_F, P_C, \bar{Q}_F) = 0 \tag{25}$$

$$\Sigma Q_C = f_C(P_F, P_C, P_L) = 0 \tag{26}$$

$$\Sigma Q_L = f_L(P_C, P_L, P_S) = 0 \tag{27}$$

$$\Sigma Q_S = f_S(P_L, P_S) = 0 \tag{28a}$$

By substituting Eq 24 into Eq 28a, Eq 28b is given as a function of $P_L$

$$f_S\{P_L, f(P_L)\} = 0 \tag{28b}$$

Thus, $P_F$, $P_C$, $P_L$, and $P_S$ can all be obtained from these formulas, and then, $\bar{Q}_F$—the required amount for smoke stopping at *L-S*—is given by solving Eq 25.

Figure 5 shows a flow chart representing the just-mentioned procedure. Further, if the corridor, lobby, or staircase is chosen as the location for the forced air supply or smoke exhaust, a calculation procedure can be composed in a similar manner.

## An Example Building and Hypothetical Conditions

Figure 6 shows a typical floor plan of a 50-story building. As this is bisymmetrical, we may calculate only half, including fire rooms, providing that the other part is under the same conditions as this part. Then, the previously introduced mathematical model and calculation procedure can be applied also to this building. Those values proposed in the previous section are adopted in principle as general conditions for this example. First, a case consisting of these conditions, as shown in Table 6, was set as the standard case. Then, in order to study the influence of such individual conditions as external wind velocity, openings, smoke tower, forced ventilation, etc., on the smoke control effect, 16 cases are chosen, and only one factor of the conditions in the standard case is changed in each of the cases (see Table 7).

## Results of Calculation for Example Building

Table 7 shows the required amount of air supply (+) or smoke exhaust (−) calculated for 12 (4 × 3) variations for each of the 16 cases, that is, 4 cases of the locations of air supply or smoke exhaust and 3 cases of the locations at which the smoke is stopped. When a smoke stopping position is situated somewhere between the fire room and some forced ventilating

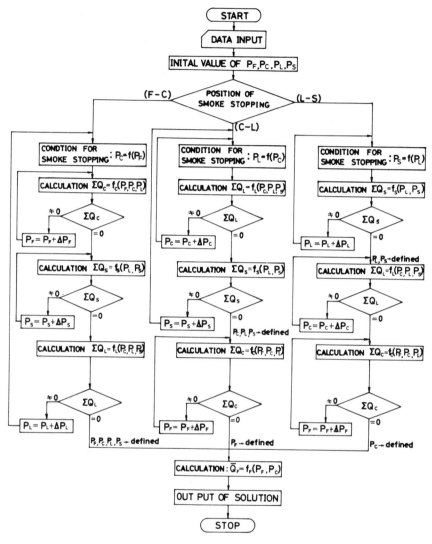

FIG. 5—*Procedure of a simplified calculation method (in the case of forced ventilation at fire room).*

space, the required smoke stopping must be achieved by air supply. On the other hand, when a forced ventilating space is situated somewhere between the fire room and some smoke stopping position, it must be achieved by smoke exhaust. Therefore, the sign on the figures in the table is generally plus for the former and minus for the latter. In opposite cases, it means that the aim or the smoke stopping is achievable without forced ventilation at the position proposed.

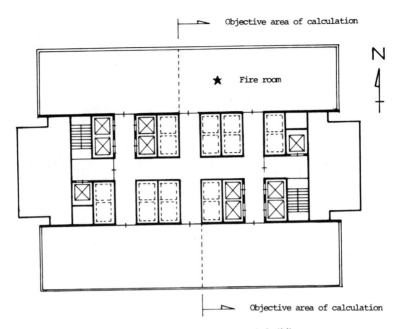

FIG. 6—*Typical plan of an example building.*

By this calculation, of course, we can obtain not only flow quantities at openings and flowpaths but also static pressures in the spaces concerned. Figure 7 gives an example showing the computed results for the mass rate and the static pressure in Case 3 under conditions where air is supplied at the lobby (*L*), and the smoke is stopped at the door between *C* and *L* (*C-L*). This figure presents a simplified view of the air or smoke flows shown by the arrows for the fire floor and every tenth floor. The values of the pressures in Fig. 7 represent the difference from the static pressure of the external air at the level of the fire floor.

Since the calculation just introduced is conducted on a particular model of a building with some limited conditions and assumptions, it is difficult to derive some general applicability about the effects and efficiencies of smoke control from these results. However, there will be remarkable difference in the required amount of ventilation, according to the location of the ventilation, even though the smoke stopping is to be done at the same location, and it is more effective to conduct the forced ventilation in the corridor or lobby rather than in the fire room or staircase, under conditions such as external windows being largely broken up on opening. For reference, the Japanese Building Code requires, in principle, provision of some systems for exhausting smoke from fire compartments.

The type of calculation methods introduced through an example are therefore extremely useful for design or evaluation of systems or measures for smoke control because, in carrying out such design, one would normally

TABLE 6—*Standard condition for comparison of smoke control effect under various conditions.*

| General Conditions | |
|---|---|
| Condition | Specification |
| Season of fire outbreak | winter |
| External weather condition | |
| temperature | −1.5°C |
| wind direction | N |
| wind velocity | 9.0 m/s |
| Number of floors of building | 50 |
| Floor height | 3.5 m |
| Position of fire outbreak | room on 2nd floor |
| Temperature | |
| fire room | 900 (°C) |
| corridor | 20 |
| lobby | 15 |
| staircase | 15 |
| duct, shaft | 10 ~ 30 |

Opening Conditions (not only for the "standard case")

| Situation of Opening | Sizes of Opening, m | | |
|---|---|---|---|
| | Height | Breadth | $H_l$[a] |
| Fire room | | | |
| window | 1.8 | 12.0 | 0.7 |
| natural ventilation opening | 0.2 | 12.0 | 2.5 |
| Fire floor, door | | | |
| F-C | 2.1 | 0.9 | 0.0 |
| C-L | 2.1 | 1.0 | 0.0 |
| L-S | 2.1 | 0.5 | 0.0 |
| General floor, window | | | |
| staircase | 1.8 | 1.0 | 0.7 |
| lobby | 1.8 | 1.0 | 0.7 |
| corridor | 1.8 | 1.0 | 0.7 |
| office room | $0.9 \times 10^{-3} \times 75.0$[b] | | 1.8 |
| Staircase, door | | | |
| ground floor | 2.0 | 1.0 | 0.0 |
| roof floor | 2.0 | 1.0 | 0.0 |
| general floor | 2.0 | $3. \times 10^{-3}$ | 0.0 |
| Opening of smoke tower | 1.0 | 1.0 | 2.0 |
| Elevator door | | | |
| to lobby | 2.0 | 0.1[c] | 0.0 |
| to corridor | 2.0 | 0.25[c] | 0.0 |
| Air conditioning duct system | $R^d = 0.2$ s²/(kg m²) | | |

[a] $H_l$ = vertical distance from floor surface to bottom of the opening.
[b] Total amount of window crevices every floor (equivalent breadth × length).
[c] Total breadth of door crevices every floor.
[d] Flow resistance of subsystem of ducts between room and main vertical shaft.

TABLE 7—Required amount of ventilation for smoke control under various conditions.

| No. | Cases, Changed Specification to "Standard Case" | | | Fire Room (F) Position of Smoke Stop | | | Corridor (c) Position of Smoke Stop | | | Lobby (L) Position of Smoke Stop | | | Staircase (S) Position of Smoke Stop | | |
|---|---|---|---|---|---|---|---|---|---|---|---|---|---|---|---|
| | | | | F-C | C-L | L-S | F-C | C-L | L-S | F-C | C-L | L-S | F-C | C-L | L-S |
| 0 | Standard case | | | -364.3[a] | -321.5 | -261.7 | 21.0[b] | -15.9 | -8.7 | 23.7 | 16.7 | -7.6 | 71.1 | 51.0 | 33.6 |
| 1 | Wind velocity | 0 m/s | | -318.6 | -269.6 | -196.8 | 17.8 | -13.7 | -6.1 | 20.1 | 13.7 | -5.5 | 61.0 | 40.2 | 15.4 |
| 2 | Window of fire room | without breakage | | -6.0 | 0.2[c] | 7.4[c] | 13.6 | 6.2[c] | 6.9[c] | 15.5 | -0.2[c] | 6.5[c] | 47.3 | -0.7[c] | -22.2[c] |
| 3 | Door of stair case at roof floor | closed | | -339.0 | -257.5 | -151.3 | .18.2 | -12.1 | -3.6 | 21.1 | 13.4 | -3.3 | 45.3 | 29.3 | 14.6 |
| 4 | Door of stair case at ground floor | closed | | -391.0 | -376.8 | -338.7 | 23.0 | 20.2[c] | -14.0 | 25.8 | 19.0 | -11.7 | 53.0 | 39.6 | 28.9 |
| 5 | Window of corridor | opened | | -258.3 | -416.8 | -365.1 | 34.0 | -38.4 | 27.3[c] | 40.3 | 14.8 | -14.9 | 124.1 | 44.3 | 23.6 |
| 6 | Window of lobby | opened | | -278.9 | -214.3 | -440.5 | 24.1 | -9.8 | -32.2 | 40.6 | 30.1 | -26.6 | 132.0 | 94.9 | 24.3 |
| 7 | Natural ventilation opening of fire room | opened | | -160.4 | -137.5 | -105.8 | 20.1 | -15.2 | -8.0 | 22.7 | 15.9 | -7.0 | 68.4 | 48.3 | 25.9 |
| 8 | Smoke tower opening | (F) | opened | -337.3 | -331.5 | -266.1 | 20.7 | -19.2 | -10.2 | 23.3 | 16.7 | -8.5 | 69.8 | 50.9 | 34.4 |
| 9 | | (C) | opened | -325.1 | -341.3 | -255.0 | 32.6 | -21.1 | -7.9 | 38.3 | 13.1 | -5.7 | 117.7 | 37.4 | 9.3 |
| 10 | | (L) | opened | -348.1 | -345.2 | -229.6 | 25.6 | -19.9 | -5.9 | 32.9 | 25.2 | -4.1 | 102.9 | 77.2 | 10.8 |
| 11 | Amount of air supply, kg/s | (F) | -10 | -354.3 | -311.5 | -251.7 | 20.1 | -15.8 | -8.6 | 23.6 | 16.6 | -7.6 | 70.7 | 50.6 | 33.1 |
| 12 | | (C) | +10 | -458.2 | -251.5 | 85.0[c] | 31.0 | -5.9 | 1.3[c] | 36.3 | 11.3 | 1.3[c] | 111.1 | 24.5 | -4.5[c] |
| 13 | | (C) | -10 | -293.8 | -372.8 | -298.7 | 11.0 | -25.9 | -18.7 | 12.7 | 18.2 | -14.7 | 43.8 | 55.9 | 41.9 |
| 14 | | (L) | +10 | -442.0 | -453.8 | 120.3[c] | 27.7 | -25.4 | 3.1[c] | 33.7 | 26.7 | 2.4[c] | 105.7 | 82.2 | -10.1[c] |
| 15 | | (L) | -10 | -308.2 | -238.4 | -319.4 | 12.6 | -11.1 | -19.9 | 13.7 | 6.7 | -17.6 | 47.2 | 33.5 | 44.8 |
| 16 | | (S) | +10 | -347.7 | -281.8 | -201.7 | 19.4 | -13.4 | -5.7 | 22.0 | 14.7 | -5.1 | 61.1 | 40.9 | 23.6 |

Required Amount of Ventilation, kg/s — Position for Forced Ventilation

[a] Negative values mean amount of smoke exhaust.
[b] Positive values mean amount of air supply.
[c] The case in which the objective control will be possible even if the contrary ventilation should be performed up to the amount represented by the underlined figure.

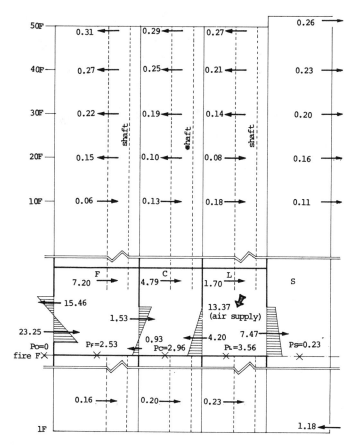

FIG. 7—*Flow quantity (kg/s) of air or smoke through each opening at fire floor and every ten floors, and static pressure (kg/s) on fire floor (for Case 3, forced air supply at lobby, and smoke stopping at the door between C and L).*

be required to carry out a great number of calculations of different conditions for even one building.

*H. J. Roux*[1]

# The Role of Tests in Defining Fire Hazard: A Concept

**REFERENCE:** Roux, H. J., "**The Role of Tests in Defining Fire Hazard: A Concept,**" *Fire Standards and Safety, ASTM STP 614,* A. F. Robertson, Ed., American Society for Testing and Materials, 1977, pp. 194–205.

**ABSTRACT:** Recent increased concern about life safety by various interested public agencies has led to questioning the credibility of small-scale fire tests in predicting the performance of a product or system in a real-fire situation.

It is proposed that many of these fire test methods, though, can be upgraded to fire-characteristic test methods by the application of appropriate criteria. Furthermore, the results from one or more fire-characteristic test methods, plus other parameters, could then be integrated to establish the potential for harm (PH) of a product or a system. The fire hazard of a product or system can then be established by combining the potential for harm with the degree of exposure.

One pragmatic approach to establishing the integration equation for the PH of a product or system is to start with the available fire statistics for that product or system and to prepare a scenario that describes the involvement and the effects of the subject item in a fire situation. Then an integration equation that fits this scenario can be written. This approach would identify the fire-characteristic test methods and parameters that are needed to assess the PH of that product or system. Included in this concept is the idea of using full-scale fire tests either to prove the integration equation for a given product or system or to assess directly the PH of that given product or system.

Alternatively, full-scale fire tests can be used to derive the integration equation for a given product or system when no fire statistics are available for that product or system.

**KEY WORDS:** fires, fire hazards, tests

Our concern for the hazard of living has many facets, not the least of which is the fear of exposure to an unwanted fire. We limit this hazard by prevention or control of the unwanted fire or absentation. Yet, our minds are usually free of any thought of fire hazard. It is only the headline reporting a multiple-death fire or the experience of an immediate event that awakens our thoughts to this concern.

[1] Coordinating manager, Product Fire Performance, Armstrong Cork Co., Lancaster, Pa. 17604.

Chauncey Starr explained this type of response in an earlier analysis [1][2] of human activities. He divided all human activities into two categories: those in which the individual participates on a voluntary basis and those in which the participation is involuntary. We are willing to ski, for example, because the hazard involved in skiing is equal to or less than the hazard we voluntarily accept for the value received by skiing (Fig. 1). In a different situation, we are willing to use electric power in our homes because the hazard involved is equal to or less than the hazard we involuntarily accept for the value received. Accordingly, based on this thesis, fire on a national basis is also an acceptable hazard. There is no great, sustained, public outcry about the 12 000 deaths due to fire year in this country. As shown, plotted by the author on Starr's diagram (Fig. 1), fire hazard is much less than the hazard that society involuntarily accepts, irrespective of the value received.

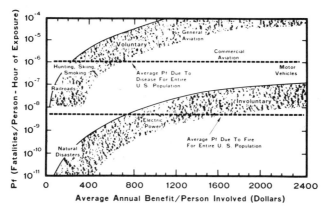

FIG. 1—*Starr's diagram.*

There are two important conclusions that can be drawn by the application of this analysis to fire hazard:

1. Hazard involves some degree of exposure. In other words, there is no hazard in skiing to the person who does not ski. Similarly, absentation from an unwanted fire eliminates the hazard.

2. The degree of hazard is a direct function of the degree of exposure. The low value of fire hazard plotted on Starr's diagram reflects the low degree of exposure of a person in the United States to an unwanted fire. Take into consideration the number of fires that do occur, specifically those that result in death, versus the number of hours of exposure for the entire population (some 200 000 000 people, 24 h/day, 365 days/year). Keep in mind that the number of fires of this type that occur in selected

[2] The italic numbers in brackets refer to the list of references appended to this paper.

situations, such as an unregulated nursing home, and the number of hours of exposure for those persons in such an environment, results in different degrees of exposure and different values of fire hazard than are plotted on Starr's diagram. This is of particular significance for the person in a high-exposure situation, although he is statistically counted as part of the entire population.

It is also interesting to note from this analysis that the prevention and control of fire hazard does not need to be 100 percent successful; perfection in this area is not a requirement of society. Unfortunately, though, there is a tendency among some researchers and some fire-protection practitioners to continue, *ad infinitum,* consideration and discussion of all of the variables, for example, different fire situations that might apply to defining a given fire hazard. There is clearly a greater need for immediate action to solve a substantial part of the problem, although not all of the problem is solvable or even understood. This paper is presented with this thought in mind.

**Definition of Fire Hazard**

In order to properly address the fire-hazard problem, we must remove the confusion that exists in defining and understanding fire hazard itself, particularly the word "hazard." It is thought that one can test for fire hazard; however, this is not altogether true when one must take into account people exposure. It may be true for property exposure when the expense is justified. However, in most cases, the loss in a real fire is the only positive fire-hazard test result available, and obviously it is "after the fact."

More specifically then, fire hazard (FH) of a product or system, or of a room or space, or of a situation is the combination of the potential for harm (PH) of that subject and the degree of exposure (E) of people or property to a fire involving that subject. Qualitatively, this is expressed as follows

$$FH = PH \times E$$

It is self-evident from this relationship that there is no fire hazard if there is no exposure!

To digress for a moment, the selection of the expression "potential for harm" is justified from several viewpoints. First, it is self-explanatory, and it clearly and accurately delineates the value which it reports. Second, it is not easily confused with the words "hazard" or "risk," which, by definition, include the factor of exposure. Third, it is not alarming, in the sense of disturbing or frightening the user, but rather serves as a warning in the sense of alerting or cautioning the user, a much-preferred connotation. In addition, there is a scaler implication in this expression that is recognizable in its acronym, PH, in respect to the more familiar acidity scale of pH.

It is also the intent of this paper to emphasize the fire hazard of a product or system. Fires no longer destroy cities, or many buildings completely, with the possible exception of unregulated single-family homes or comparable construction. The principal area of concern is a room or space and the products and systems within that space. *A priori*, the fire hazard of a material is denied since a material exists only as part of a product or system in a real-fire situation.

The fire hazard of a situation, by definition, is the fire hazard of a given room with a given product added to that room. Therefore, the fire hazard of a product is the change in the fire hazard of a situation as caused by the addition of the product to the existing room or current situation. The fire hazard of a product is generally not directly measurable, even assuming a given degree of exposure, because a product does not act independently of the room (environment) in which it is located. One exception to this premise (and there may be others) is the clothing on a person, and this is quite a limiting situation.

Of greater importance, though, is the proposition that measuring fire hazard directly is improbable, if not impossible, when a degree of exposure of people is included. True, it is possible to include a degree of exposure of property in a fire-hazard test of a room or situation (which indirectly could lead to a measurement of the fire hazard of a product), but the expense and complexity of the test makes this improbable. As a matter of consideration, if one considers all avenues, it is even possible to simulate a degree of exposure of people.

Nonetheless, it is definitely possible to measure the PH of a room, since it is independent of exposure, and this is typically what is measured in full-scale (room) fire testing. Criteria for this mode of testing are now being established [2]. As mentioned before, it is also possible to measure the PH of a product by the difference in the PH of a room with and without the product in question. Practically speaking, though, there is a strong desire to measure the PH of a product directly without the additional expense and encumbrance (complexity, limited testing facilities) of two PH room tests, especially for an individual product.

## Measurement of Potential for Harm

A procedure for measuring the PH of a product directly, albeit by induction, is within reach (Fig. 2). This procedure makes use of an integration equation to bring together the results from one or more fire-characteristic test methods (FC), plus other parameters (OP), in an appropriate

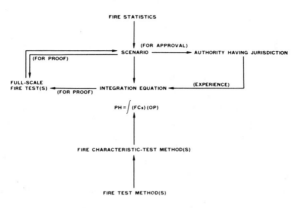

FIG. 2—*Product potential for harm.*

order that generates (induces) the PH of the product in question. It is expected that each type of product would have a different integration equation. The general form of this equation, in any event, is as follows

$$PH = \int FC_1{}^a \times FC_2{}^b \times FC_n{}^x \times OP_1{}^a \times OP_n{}^x$$

Integration equations for this type of purpose are not new [3–5]. As a matter of fact, most authorities having jurisdiction integrate, however subconsciously, the results (real or imaginary) from one or more fire-characteristic tests. Factors, such as ease of ignition and rate of flame spread, together with information on other parameters, such as product size and use (environment), are used to arrive at an estimate of the PH of the product. This is then used, in combination with an estimate of the degree of exposure, for product approval.

(For the future, there is also the thought of a supraintegration equation that brings together the results from several product-integration equations, with other parameters, in an appropriate order to generate the PH of a room. This is more properly the jurisdiction of a building code.)

The integration equation can be derived in several ways for a given type of product. One way is to utilize directly the experience of the authorities having jurisdiction, formalized if desired by the Delphi technique, to write the integration equation. Another way is to start with the fire statistics, for the type of product in question, and to prepare a scenario that describes how the product typically becomes involved in and contributes to a fire. After approval [6] of this scenario, it is translated into the integration equation.

By way of enlightenment, the scenario from this source is virtually a word picture of the fire hazard of the product since it includes a complete account of both the product's potential for harm (used to derive the integration equation) and the degree of exposure. The specific information included in a typical scenario [7] consists of the type of loss (death, injury, or property damage, or all of those), type of occupancy, time of fire, ignition source, spreading agent, and direct cause of loss (smoke or gas or heat or flame or all of those). In prospect, the concern in most scenarios is to the length of time before flashover occurs in the room in which the product is located.

In the absence of fire statistics for a given type of product, the full-scale (room) fire test is an alternate source for the scenario needed to develop the integration equation. Two methods can be used to develop this source: (a) a single full-scale (room) fire test, which can be used to confirm the scenario prepared from fire statistics, or (b) a pair of fire tests, one with and one without the product in question, with the difference in results used in the scenario preparation.

An excellent example of the use of a scenario to prepare a valid integration equation for a given type of product is the mattress-flammability standard [8]. Although fire statistics were used by the National Bureau of Standards to prepare the original scenario, a final scenario with sufficient detail was possible only after full-scale (room) fire tests of mattresses. The resultant integration equation is, by the way, very simple; it contains only a single fire-characteristic test (ease of ignition by a burning cigarette)!

One study is now underway on the mechanics (mathematics) of writing integration equations for various types of products, and early results [9] have identified some of the problems involved. For instance, information is lacking on the component limits (heat, flame, smoke, gas) of the human-response envelope. This information is needed to set the appropriate order of the various results of fire-characteristic test methods and other parameters in the PH integration equation. This is further complicated by the need to establish the bases for the component limits, that is, temporary loss of useful function (control response) or permanent loss of life. At the same time, there is another study [10] underway that is directed toward preparing appropriate criteria for the qualification of a fire test method as a fire-characteristic test method, a key element in any integration equation.

## Fire-Characteristic Test Methods

Many fire test methods known today are not capable of being upgraded to FC test methods without substantial change. However, there is no need to qualify all fire test methods as FC test methods; many will continue to serve for other uses, such as control or quality-assurance-type tests. To these methods, though, when under the jurisdiction of the ASTM Board of Directors, the following caveat is attached:

> This standard (fire test method) should be used solely to measure and describe the properties of materials, products, or systems in response to heat and flame under controlled laboratory conditions and should not be considered or used for the description, appraisal, or regulation of the fire hazard of materials, products, or systems under actual fire conditions [11].

This caveat emphasizes the difference between performance-predictive type FC test methods and other fire test methods. It is also, in part, one reaction to the increased concern about life safety by various interested public agencies [12] that has led to the questioning of the credibility of small-scale fire tests to predict the performance of a product or system in a real-fire situation. This is the role of an FC test method.

To digress for a moment, the selection of the expression "fire-characteristic test method" is justified from one particular viewpoint. Most importantly, it explains the type of answer sought by the particular test, a subject ignored with the more common designation "fire test method" and not definitive enough with the alternate, suggested designation "fire-performance test method."

The proposed [10] criteria for the qualification of a fire test method as an FC test method are as follows:

1. The method shall measure one or more fire characteristics. Herein lies one of the problems, the proper categorization (identification) of fire characteristics. One list [10] of fire characteristics contains the following ten items:

(a) Ease of ignition.
(b) Rate of heat release.
(c) Rate of fire growth (change in rate of heat release).
(d) Rate of smoke production.
(e) Rate of toxic (and corrosive) gas production.
(f) Rate of flame spread.
(g) Rate of oxygen depletion.
(h) Rate of burning (consumption).
(i) Structural integrity.
(j) Ease of extinguishment.

Although most of the fire characteristics on this list are of the same category (a basic fire property of the product), some, such as oxygen depletion, depend upon the environment in which the product is located. Further study is, therefore, needed to confirm and define this list of fire characteristics.

2. The method shall test the product or system in the form of its use. Incorporated in this criterion are the concepts of geometry and orientation (of the product) and, conceivably, a ratio of area (of product) to

volume (of space). The latter is of particular consequence in assessing the interaction of environment on performance.

There are two ways the environment can be considered when it is not automatic, as in the case of a full-scale (room) fire test. The first way is as part of the fire-characteristic test method. The second way is as part of the integration equation for PH, as one of the other parameters in the equation. Either way, any success in correctly assessing the PH of a product or system is dependent on the skill with which the effect of the environment (room or space) is embodied in this measurement. It is fortunate that most products have a limited number of applications and, it is hoped, a commonality of environment caused by the other properties and cost of the product or the strictures contained in the instructions and recommended practices for the product.

3. The method shall provide a quantitative result. This criterion is needed in order to obtain the proper form of input to the PH integration equation. It also avoids the prejudgment inherent at this level in qualitative tests of the go/no-go type.

4. The method shall use an ignition or fire source that is related to a real-fire situation. An idealized ignition or fire source is obviously sought for test simplicity; but because of the complexity of a real fire, test ignition must be related to a real-fire situation. This relation, interestingly, is prescribed by the same scenario as that used to write the PH integration equation.

In this regard, it is expected that there will be a need for a better definition of fire exposure. This definition might use a target or a mathematical expression that "reads" radiant heat flux, convective heat flux, and gas temperature as a function of time [13]. The capability of the test method to use different sizes of ignition or fire sources is also of some consequence, particularly from the economics side of performing the test.

**Control of Fire Hazard**

Although the authority having jurisdiction would be expected, in theory, to limit the fire hazard of a situation for regulatory purposes, it is expected that he will, in practice, limit the PH of a room or the PH of a product. It is further expected that he will be guided in doing this by an estimate of the degree of exposure associated with the room or product. The degree of exposure is the probability of a person or property being exposed to the fire, coincident (the time element) with the probability of a fire occurring in the same place. In some cases—for example, property and a nonambulatory patient that are fixed in space—the degree of exposure is dependent only on the probability of a fire.

The important thing is the measurement of the PH of a room or product that the authority having jurisdiction is able to use. Coincidentally, it is of consequence to determine, particularly for a product, whether the value sought for the PH is of the mean or average (most probable) case or of the extreme (most severe) case of involvement in a fire. Guidance for this determination is sometimes available from the fire statistics for the type of product (or room) in question.

The value obtained for the PH of a room or product is without dimension and thus not subject to prejudgment. Also, the authority having jurisdiction is free to consider several different methods of use of this value. For example, with experience and the knowledge of the PH of other representative products, a limit could be set for the PH of a given type of product or different limits for different types of occupancies. This is the typical approval/disapproval type of decision.

Another method, more productive but involving a more difficult type of decision, is to accept the measured PH, regardless of its value, and then to require either enforced proper-use instructions and recommended practices or a "success-tree" [14] approach. The latter is an engineering approach to the use of other fire-protection measures ("prevention of fire, ignition" and "manage fire impact" are the major headings) in the same occupancy to obtain a given fire safety objective, although there is a patently unacceptable room or product within that occupancy.

## Conclusion

A proposed procedure has been presented in this paper that provides for the upward progression (scaling?) from a fire test method, to a fire-characteristic test method, to the calculation or test of the potential for harm, and, finally, to the measure of the fire hazard of a product or system. At this time, it is expected that appropriate fire-characteristic test methods can be prepared.

Concurrently, it is not expected as yet that any PH integration equations can be developed other than those that contain only one fire-characteristic test method or those that evenly weigh the results of all of the fire-characteristic test methods involved in a particular equation.

The need for a comprehensive procedure is easily substantiated. There is the need to know, accurately and at the least cost, the fire hazard of all of the new products and systems that enter the world each day, if not simply to provide the basis for any sort of corrective action. But there is also the need for a procedure or plan that precludes continuation of the extremely fragmented and, therefore, inefficient approach to fire testing that exists today.

## References

[1] Starr, Chauncey, Science, Vol. 165, 19 Sept. 1969, pp. 1232–1238. Copyright 1969 by the American Association for the Advancement of Science.

[2] "Recommended Practice for Room Fire Tests," Task Group IV of ASTM E-39.10.01 (Research) Subcommittee, draft.

[3] Harrington, E. C., Jr., *Industrial Quality Control,* April 1965, pp. 494–498.

[4] Minne, R., in *Ignition, Heat Release, and Noncombustibility of Materials, ASTM STP 502,* American Society for Testing and Materials, 1972, pp. 35–55.

[5] "Furniture—A Simple Approach to Reducing the Fire Hazard," *Draft of Handbook for Designers,* The British Plastics Federation, July 1975.

[6] Public Concern Division, ASTM E-39 Committee; or others (authorities having jurisdiction).

[7] "Reducing the Nation's Fire Losses, The Research Plan," National Bureau of Standards, July 1975.

[8] "Standard for the Flammability of Mattresses," DOC FF 4-72, U.S. Department of Commerce.

[9] "The Use of Fire Research Information in Assessing the Fire Hazard of a Situation; A Preliminary Listing of Research Needs," Task Group III of ASTM E-39.10.01 (Research) Subcommittee, draft.

[10] ASTM E-39.10.02 (Planning) Subcommittee.

[11] "ASTM Policy Defining Fire Hazard Standards," Adopted 18 Sept. 1973.

[12] "Disclosure Requirements and Prohibitions Concerning the Flammability of Plastics," *Federal Register,* Federal Trade Commission, 6 Aug. 1974.

[13] Castle, G. K., "The Nature of Various Fire Environments," *American Institute of Chemical Engineers Loss Prevention Symposium,* 14 Nov. 1973.

[14] "Decision Tree," National Fire Protection Association Committee on Systems Concepts for Fire Protection in Structures, Nov. 1974.

# DISCUSSION

*P. H. Thomas[1] (oral discussion)*—I go along most of the way, perhaps all the way, with the "potential for harm" side. My only question relates to the combination of this with "exposure." A feature of fire is that it is something that one can intervene in, to put in sprinklers, detectors, and all sorts of other systems, so it is, in some ways, different from some other things like being hit by a meteorite and certain risks to which people develop fatalistic attitudes. They say, "If that happens, that's the end." So I want to ask if people weight their attitude to exposure and to the consequences of that exposure differently, does it not affect compounding of these two probabilities as if they were independent? Does fire hazard depend only on some combination and the components PH and E not matter in themselves?

*H. J. Roux (author's oral closure)*—I am not sure that there is a quick answer to your question. I think the key concept in my paper, to be redundant, was the idea that hazard involves both a potential for harm and also exposure. I do recognize that, in many situations, we will not be able to either control exposure or the potential for harm because this is a symptom of the real world. We will have to accept, in many cases, a given degree of exposure. And, as such, we will deal only with the subject of

---

[1] Fire Research Station, Borehamwood, Herts, United Kingdom WD6 2BL.

potential for harm and try to reduce this as much as possible. And, consequently, I think, in many cases, the final result or fire hazard, will then be subject to the efforts of the authorities having jurisdiction.

*H. L. Malhotra*[2] *(oral discussion)*—Figure 2 gave me the impression that full-scale room tests are regarded as providing positive proof of the behavior pattern in a fire. Having designed and seen numerous full-scale tests I would like to suggest that such tests also can be subject to as much variability and some of the other drawbacks as the tests which they are replacing. I would like to suggest that we need to change our attitudes toward fire tests whether they be the bench-type microtests or the complex room-size macrotests. We should not regard tests as giving a definitive answer on the safety or hazard of certain materials or products but only as a means of providing technical information on which judgment can be based.

Fire tests by themselves are incapable of making judgments. These should be made by the users on the basis of the information the tests provide and using a systems approach in which the test information is assessed alongside other data relevant to the assessment of the fire hazard.

The unsatisfactory situation to which reference has been made in this paper, developed not only because of the inadequacy of a range of fire tests, but also because of the misuse to which they were often put. It is worth drawing attention to the activity which is taking place in the International Standards Organization (ISO). A special panel with representatives from Australia, France, Switzerland, the United Kingdom, and the United States has been discussing the philosophy of fire tests and their role. It is intended to use the final document as a guide to all ISO technical committees engaged on preparing fire test procedures. One aim of this activity is to prevent the occurrence of the unsatisfactory situation by providing a better understanding of the role and functions of fire tests as well as their limitations.

*H. J. Roux (author's oral closure)*—It is my hope that the concept presented in this paper is reflective of your own thoughts. For example, one of the criteria that is presented for a fire characteristic test method is that it shall provide a quantitative result rather than a go/no-go type of answer directly. Further, the idea of potential for harm is to force a judgment which is not based on a single test result alone, in assessing the fire hazard of a product.

*W. A. Dunlap*[3] *(oral discussion)*—Does the potential for harm factor include both probability and severity?

*H. J. Roux (author's oral closure)*—Severity would be a characteristic of the product in question, as reflected in its PH, based on the results of

---

[2] Fire Research Station, Borehamwood, Herts, United Kingdom, WD6 2BL.
[3] Codes and standards specialist, The Dow Chemical Company, Midland, Mich. 48640.

the appropriate fire characteristic test methods. For example, smoke production might be an important factor in the PH for a given type of product. Product A, though, might produce more smoke than Product B, and therefore its PH value would be higher. In regard to probability, I included it only at the interface with degree of exposure. There may be other ways of looking at it.

# Society's Response

*J. E. Bihr*[1]

# Building and Fire Codes: The Regulatory Process

**REFERENCE:** Bihr, J. E., "**Building and Fire Codes: The Regulatory Process,**" *Fire Standards and Safety, ASTM STP 614*, A. F. Robertson, Ed., American Society for Testing and Materials, 1977, pp. 209–221.

**ABSTRACT:** The complete regulatory process involves the application and enforcement of building and fire codes. Building codes govern new construction and embody standards designed to protect occupants and property from the hazards of fire. Fire codes are intended to preserve fire safety features of building codes and to govern fire safety practices involving the processes and uses within buildings. These codes must be compatible. Additional research is needed to address some of the newer fire problems now confronting regulatory officials.

**KEY WORDS:** fires, building codes, fire protection, regulations, fire safety, fire hazards

The regulatory process, as it is addressed in building and fire codes, is not fully understood. A too-frequent assumption is that matters relating safety are delineated in fire codes and enforced or applied by the fire department of a municipality. Actually, the main function of a fire department is the logistical response to actual fires and, no less important, perhaps more, the supervision of fire prevention programs and the supervision of maintenance of fire protection equipment.

Fire maintenance programs usually follow regulations detailed in either a building code or a fire prevention code. For example, the building code may require the installation of an automatic fire-extinguishing system in a building in view of the building size. The fire maintenance program is designed to ensure that the fire-extinguishing system, as installed, will be maintained in a ready condition. It is the building code which establishes the basic fire safety characteristics of a structure. It is the building code that sets out building areas and heights, fire endurance of building elements, exit design, and the installation of fire protection equipment. And it is the enforcement

---

[1] Managing director, International Conference of Building Officials, Whittier, Calif. 90601.

of a compatible fire code which ensures that the basic fire safety elements and systems built into a structure are maintained and that the processes and uses conducted within a building are consistent with recommended safe practices.

### The Regulatory Process

Although carried out at the local municipal level, the regulatory process throughout the United States, insofar as it affects private and, to a great extent, public buildings, is a derivation of police power reserved to the states. States, in turn, have delegated this in whole or in part to local municipal governments.

The building code is generally thought of as that document which regulates all building construction. It includes within its definition codes dealing with zoning, pressure vessel inspection, elevators, etc. This explanation is generally acceptable since the entity responsible for building safety also has administrative and enforcement responsibilities in these other areas. In addition, responsibilities sometimes include areas dealing with building maintenance. This is particularly true in the housing field where housing codes are adopted to guard against the growth of substandard housing or to provide a basis for rehabilitation of existing residential buildings. Fire codes, on the other hand, are strictly concerned with the maintenance and use of buildings. And, if adopted, they are generally enforced through a fire prevention agency linked to the local fire department. I say "if adopted" because many communities in the United States that are basically rural in character, such as unincorporated areas of counties, have voluntary fire departments or separate fire districts that divide a county into several areas which are not politically compatible in terms of enforcement of fire codes. Yet, a code which deals with the continued fire safety maintenance of a building has a significant impact on the fire record, and fire code adoptions should and must be encouraged if we are to make continued advances in the protection of life and property from the ravages of fire.

Building and fire codes, although different in their approach (that is, new construction versus maintenance), must be wholly compatible. It would be ludicrous from an administrative viewpoint to require exists for assembly occupancies to be maintained in accordance with rules and regulations that are in conflict with the exit situations which were originally designed into the structure on the basis of the building code. The Uniform Building Code[2] and Uniform Fire Code[3] are examples of wholly compatible documents which separate construction and maintenance yet utilize compatible and consistent terminology in requirements to ensure that the fire safety features and designs implanted in buildings are adequately maintained.

[2] Uniform Building Code, International Conference of Building Officials, 1976 ed.
[3] Uniform Fire Code, International Conference of Building Officials and Western Fire Chiefs Association, 1976 ed.

## Model Code Influence

A 1971 survey[4] of building code use in the United States indicates that 3 percent of cities with a population larger than 10 000 have no building code regulations. The survey went on to report that approximately two thirds of all local codes are patterned after one of the four model codes. The accelerated adoption of model codes over the last few years would indicate that these ratios have been further enhanced so that the model codes are the dominant factor in the United States' current regulatory process. These codes and their publishers are as follows:

1. Basic Building Code—Building Officials and Code Administrators (BOCA) International, Inc.
2. National Building Code—American Insurance Association.
3. Southern Standard Building Code—Southern Building Code Congress, International (SBCC).
4. Uniform Building Code—International Conference of Building Officials (ICBO).

The code change processes of BOCA, ICBO, and SBCC provide an open forum for exchange of views, and the resulting code publications of all groups spread the cost of their development (several hundred thousands of dollars) among all the users. The model code organizations provide additional valuable services that are vitally necessary in the enforcement programs of adopting jurisdictions.

New building products and systems are reviewed monthly by ICBO, for example. Those meeting the performance standards of the code are communicated to all member jurisdictions through the medium of a descriptive report that identifies the product or system and its uses and limitations. The professional staff assembled is able to serve hundreds of communities in a uniform fashion. Cities acting alone would have to provide identical levels of service. The large monetary savings of this joint activity are obvious.

The educational program of ICBO is another broad and diversified service that provides communities with training courses, seminars, educational aids, and an inspector certification program. The dynamics of this building industry demand that those engaged in the regulatory field be absolutely current. To achieve this individually would represent a cost that even our largest cities would find difficult to meet.

Maintenance of a professionally trained staff of civil and structural engineers to assist in code interpretations and plan review of major structures is yet another highly utilized service and one sorely needed in smaller jurisdictions.

Although these descriptions are in terms of ICBO, BOCA and SBCC also offer services of varying character in these areas. The significant point is

---

[4] Applefield, Milton, "Use of Fire Limits in the United States," United States Forest Service, 3 Nov. 1971.

that these agencies not only exist, but they offer the valuable asset of uniformity in a most effective and economical way.

## Building Code Characteristics

Model building codes classify building occupancies and uses into groups that are similar in nature with respect to the fire and life hazards that buildings pose. These codes are all framed to permit a wide range of uses within a given building occupancy. Once a certificate of occupancy is issued, the building owner or lessee is free to conduct many diverse activities which, under current code structure, do not change the basic character of the occupancy. The Uniform Building Code, for example, groups together such uses as wholesale and retail stores, office buildings, factories and workshops (not using highly flammable or combustible materials), and storage or sales rooms for combustible goods.

One purpose of such grouping is to include those uses which reflect a similar hazard to life based on combustible content or fire load. This approach also benefits the owner, since he is free to consider a wide range of potential building uses over the projected or amortized life of the structure. This point of variable occupancy is stressed because wide-ranging uses, which involve corresponding broad changes in type of contents, should be a fundamental consideration when deciding what part of the unchanging building structure is to be regulated, as well as the manner in which the regulations are to be applied. There is no doubt that there are situations where the character of the occupancy can be rigidly controlled, particularly at the owner's option. Examples include military installations and certain other governmental buildings. However, for the great majority of construction, such controls are presently outside the scope of normal municipal activities. Since there is minimal content control, it is necessary to consider this variable in developing regulations.

Building construction types are classified by building codes into categories resembling varying degrees of fire endurance or fire resistance. Areas and heights are assigned to the different occupancy categories on the basis of the different building construction type in which they are housed, the greater height and area enjoyed by the most fire-resistive building type and so on down the scale. The assigned height and area and the construction type stem from studies of fire loading and occupancy loading as well as spatial separations of a building from its neighbors. Where vacant usable adjacent land occurs, the codes assume that the land is occupied or, in other words, that the building in question will face the greatest potential hazards. Construction types of fire-resistive character have specified hourly fire-resistive ratings assigned to the principal supporting elements. And, multi-story buildings are required to achieve a great degree of compartmentation to guard against the story-to-story transfer of fire. In addition to floor-to-floor separation, compartmentation also embraces the enclosure of vertical openings and concealed spaces which have a potential for contribution to fire spread.

There are additional refinements dealing with the hazards of fire spread. For example, occupancies or uses that are incompatible are required to be separated from each other in an envelope manner. Buildings with excessively large areas must be subdivided by area separation walls with sufficient fire endurance to constitute separate structures. These walls may vary from 2 h for buildings of low fire-resistive character to 4 h for buildings of the highest type. Wherever fire endurance of building elements is prescribed, it is based upon demonstration of performance under ASTM Fire Tests of Building Construction and Materials (Including Tentative Revision) (E 119-73). Materials and systems of materials that have achieved ratings derived from this standard ASTM test are listed in substantial detail in building codes, in supplemental pamphlets, or in listings published by agencies such as Underwriters Laboratories. Sufficient information is provided to ensure that the construction, as built, will have the same integrity as the fire test specimen. This requires careful analysis and a degree of ingenuity since the as-built construction varies significantly in size from the tested prototype.

The flammability of interior finishes and their contributions to fire spread are addressed in codes. The ASTM Test for Surface Burning Characteristics of Building Materials (E 84-73) (tunnel test) is the primary code tool which assesses the relative surface flammability of building finishes. This test procedure is also utilized as an indication of comparative smoke density of the products of combustion that may be expected from these finishes. Plastic materials utilized in the building structure are approached in a similar manner as interior finishes but may be further restricted (as in the Uniform Building Code) on the extent of use in given situations, primarily to glazing functions.

The codes utilize the ASTM Test for Noncombustibility of Elementary Materials (E 136-73) as a basis for categorizing the combustibility of building materials. It is also necessary to employ concepts involving potential heat and heat release along with tests for noncombustibility to properly evaluate some of the newer building products. All of the model codes are performance oriented, so it is possible to develop approaches which will accept innovations without affecting fire safety.

Automatic fire-extinguishing systems are given responsible assignments in the codes. They are the means, for instance, whereby basic building areas can be multiplied. In the Uniform Building Code, for example, for one-story structures, the areas can be increased threefold, while in a multistory building, areas may be doubled, assuming fire-extinguishing systems, such as fire sprinklers, are installed throughout the building. These systems are also required to address specific hazards. They must be installed in all retail sales buildings with floor areas in excess of 12 000 feet$^2$, as well as in assembly areas having exhibit spaces of 12 000 feet² or more. Fire-extinguishing systems have always been required traditionally in the areas that present difficult or impossible access from a fire-fighting standpoint, such as in basements and cellars or in windowless structures. They are also a suitable alternative to certain compartmentation requirements in high-rise buildings.

The use of smoke detectors in gaining in codes. A major trend is toward mandating their installation in new residential buildings in areas adjacent to sleeping rooms. Dependent upon the building design, several units may have to be installed. They also are used in some codes as a suitable actuating mechanism for fire doors protecting vertical shafts or providing envelope protection in corridors. Similar requirements find their use in the air-conditioning, heating, and ventilating systems of buildings.

Another fire protection tool utilized by codes for buildings of extensive area is roof venting. This facilitates the removal of both heat and smoke in the event of fire, building roofs being divided into large but separate areas and vented for this purpose.

The relationship of fire hazard to occupant safety is, of course, a paramount concern of building codes. In low buildings and in buildings where persons are ambulatory, codes prescribe appropriate parameters to ensure adequate widths of corridors, stairways, and doorways. This is accomplished by determining minimum widths for a single person exiting and then developing minimum standards for the component portions of the exit path. The basic parameter for sizing exits is the occupant density for various uses. The number and width of exits is a function of the total occupant load determined by this density factor. Minimum exit dimensions accommodate a substantial number of occupants so that the actual occupant load only becomes a governing factor for the larger occupancies. For example, the Uniform Building Code requires two exits from a retail sales room when the occupant load exceeds 50. The minimum width of an exit door is 3 ft. Since the minimum width in feet is defined as the occupant load divided by 50, a 3-ft-wide exit doorway could accommodate 150 occupants.

When more than one exit is required, they must be arranged so that if one becomes blocked, the others will be available. The two-exit situation under this concept is resolved in the Uniform Building Code by requiring that the two exits be placed a distance apart equal to not less than one half of the length of the overall diagonal dimension of the area to be served measured in a straight line between exits. This technique is easily applied to irregularly shaped buildings as well as those of conventional configurations.

The number of exits are also a function of the height of a building with two exits being required when the occupant load exceeds ten. Story-to-story exiting is determined on the basis of traffic flow, considering that all occupants will not reach the desired stairway at the same time. This is done in a proportional manner, recognizing the fact that the number of exits from any one story cannot be less than the occupant load dictated by that story.

Another aspect of exit design is the length of exit travel. Excessive travel length inhibits orderly and fast evacuation. The occupant is also given the benefit of protection of an enclosed stairway or corridor. Special exit provisions are assigned to occupancies that merit special consideration, such as hospitals, where exiting is accomplished in part by attendants moving patients to areas of refuge within buildings, usually on the same floor.

**Fire Code Characteristics**

Although fire codes contain similar concepts and aims as are found in building codes, they only serve effectively when there is compatibility with the building code, and both have complementary objectives.

In a workable enforcement program, fire codes are directed towards maintenance of fire protection features regulated by building codes and the regulation of the processes that take place in buildings. This includes, for example, the safe handling of hazardous materials, fueling and defueling of aircraft, safety practices in welding and cutting operations, installation and maintenance of combustible dust collectors, maintenance of exit ways in buildings, and so on. The list seems endless, but the objective is clear. When the building is safely constructed and open for occupancy, the continued safe maintenance is all important to achieve fire safety.

The code interaction I describe is the view from the enforcement end. As our country matures, and certainly as the cost of new building construction increases, it is most important to adjust regulatory practices to conserve and maintain the inventory of existing construction.

**Code Enforcement**

The regulatory framework necessary to implement codes must be emphasized. The most well-conceived and highly principled code document will be of no value unless measures are adopted to carry out the enforcement. This goes beyond mandating the requirements into law. The adopting jurisdiction must provide the necessary staff to carry out a vigorous enforcement program. For new construction, this means a professional engineering and inspection staff with appropriate administrative and management control. New construction enforcement programs can only begin by a detailed review of plans in their entirety to verify that the project, when constructed, will meet the requirements of the code. Plan review embraces both structural and nonstructural aspects of the code and requires a high degree of professionalism, one no less than that of the persons who prepared the plans. If anything, a greater degree of expertise is needed. If plans are found wanting, then revisions must be made and a further review undertaken to verify that errors have been corrected. Jurisdictions that instigate preliminary plan reviews for major projects have provided an excellent service to forestall the occurrence of major problems. This process demands that jurisdictions involved develop a firm internal working relationship between the building and fire departments and that they adopt current and enforceable codes.

Once the plans have been approved, careful field inspection is necessary to ensure that the construction, as contemplated by the plans, is carried forward. It is futile to assume that the field inspector is going to both inspect the structure and review the plans at the same time. Inspectors must be adequately trained in field techniques and be aware of any changes or deviations to the approved plans. An in-house training program maintained on a continuing basis is necessary to verify correctness and to refresh inspection personnel and plan review personnel with construction systems and techniques that may be new or infrequent. When the building has been constructed, and a certificate has been issued permitting occupancy, provisions should be made for periodic maintenance inspections to ensure that the fire safety features and the fire protection devices that were originally installed have been properly maintained in an operable condition. An important ancillary aspect includes fire training and fire drills for the occupants of all buildings.

It is frightening that many communities do not maintain an orderly fire prevention program. Locked exit doors are the most common result of this indifference and invariably result in high life loss. The use of stairwells and corridors for storage is another example of a fire hazard that encourages disasters and could be averted by an inspection program. The failure of automatic fire extinguishers to function when needed is a further example of a lack of preventive maintenance.

**Fire Hazard Assessment**

Over the last few years, much concern has been expressed over the ability of ASTM fire test methods to adequately assess fire hazards. Practically, these methods were not developed for the purpose of making such assessments alone. Rather, they were intended to be used, and are used, as part of the data base upon which hazard assessments will be made. Clarifying language is being introduced in many of these methods to underscore their limitations, and this will be a definite aid to those who might misinterpret or abuse their intent.

Fire hazard assessment is accomplished when the results from basic research and performance tests are coupled with other parameters to deal with the potential for fire and the probability of its occurrence. Conditions vary if the fire hazard assessment is made in terms of property protection as opposed to protection of life. Often fire hazards can be mitigated through the application of techniques not measured by standardized fire tests such as the use of automatic fire-extinguishing systems or the insertion of early detection devices with provisions for rapid occupant evacuation. These various alternatives and techniques are reflected in building and fire codes which, in reality, become a basis for the systems analysis of a fire hazard.

The foregoing explanation of codes was intended to show some of the relationships applied by codes in their analysis. The significance and degree of accuracy of tests must be viewed within the perspective of code use and enforcement when judging building hazards.

Recently enacted legislation establishing the National Fire Prevention and Control Administration and the Consumer Product Safety Commission can, if wisely administered, complement the objectives of building and fire codes. An accurate definition of the fire problem is essential to its assessment and to its solutions. And, certainly, there is a need to examine and control building contents to the point of excluding materials that represent serious problems.

## Some Concluding Observations

Compatible building and fire codes enforced together are a comprehensive approach to the assessment and solution of fire hazards posed to both life and property.

Codes and the fire test methods they embody will constantly change through the years because the materials available for construction and the methods of construction change. Further, the attitudes of people change as to what codes should do and what constitutes reasonable safety within limits tolerable to the public.

Basic fire research will always be needed to respond to the changing aspects of the fire problem. A more demanding economy requires design refinements. New construction techniques must be evaluated, and their economy and effectiveness must be measured. The concept of restraint in fire endurance must be more precisely quantified. Analytical approaches should be developed, since they hold the greatest promise for economical design against fire. The fire hazard characteristics of newer materials need to be identified and judged.

The research must address the real world where building contents and building structure must be viewed concurrently. To assume that building occupants can endure a hostile fire producing salubrious gases seems a trifle farfetched. Research indeed must have a practical orientation.

In the creation of fire safety standards, it must be kept in mind that the American people do not live in complete safety; rather, they live a relatively safe life, considering the hazards that surround them. Absolute safety is not attainable. Regulations and standards cannot be so exacting, impractical, and uneconomical as to curtail construction. Requirements so restraining would find that the people affected might well prefer to be less safe with more freedom.

Many current fire deaths are attributable to fire spread in older buildings that do not meet present fire safety standards. That existing building and fire codes are not vigorously applied to new and particularly old buildings (where fire deaths and losses are greatest) may be of much greater significance in assessing the fire problem than the precision of some test methods.

### Bibliography

*Fire Protection Through Modern Building Codes,* 4th ed., American Iron and Steel Institute, 1971.

*Building Practice for Disaster Mitigation,* Building Science Series 46, U.S. Department of Commerce, National Bureau of Standards, Feb. 1973.

*Symposium on Ignition, Heat Release and Noncombustibility of Materials, ASTM STP 464,* American Society for Testing and Materials, 1969.

*America Burning,* The Report of the National Commission on Fire Prevention and Control, May 1973.

*High-Rise Building Fires and Fire Safety,* NFPA No. SPP-18.

*Fire Journal,* various issues, issued bimonthly, National Fire Protection Association (description of actual fires).

# DISCUSSION

---

*W. A. Dunlap*[1] (*oral discussion*)—Can you make an estimate of what percentage of residential construction today would be built in areas that would require early detection devices.

*J. E. Bihr* (*author's oral closure*)—The Uniform Building Code (UBC) required early detection devices in dwellings with the production of the 1973 edition. So, for the last three years, with few exceptions, all new dwellings constructed under the UBC have enjoyed the installation of detectors. The exceptions are of isolated incidences where a few local communities have opted out.

*J. R. Gaskill*[2] (*oral discussion*)—In view of your statement that contents remain unregulated, a good building official has one of its arms tied. Would you see any advantage in proposing some kind of a recommended practice or an addition to a building code to regulate the contents of public buildings? How about a recommended practice which could be broadcast for private buildings as a research tool or as a useful adjunct to the various model codes?

*J. E. Bihr* (*author's oral closure*)—A recommended practice is always

---

[1] Codes and standards specialist, The Dow Chemical Company, Midland, Mich. 48640.
[2] Fire protection engineer, Lawrence Livermore Laboratory, University of California, Livermore, Calif. 94550.

useful. The problem content control poses is the enforcement problem. How are you going to ensure that these regulations are enforced? The simplest way to enforce content regulations seems to me to be at the point of manufacture. If that could be obtained, it would solve a lot of problems. If you wait until a chair, for example, is inserted into the room, then you are very far down the road. But I believe that recommended practices are in order. They don't involve an enforcement responsibility, yet they indicate the direction in which one might choose to go.

*D. Hammerman*[3] (*oral discussion*)—There was a comment which is supported by many people that fire codes should be for the maintenance and use of buildings. I deal, to a great extent, in building construction from both standpoints, and, as we all know, in building codes, there is a great deal of fire protection. For example, height and area limits, construction classifications, fire resistance of building materials, interior finish, fire stopping, fire extinguishing systems, fire detection systems, entire chapters on means of egree, etc. My question is this: suppose (and many people will not agree with this) all this collected fire related material were placed in a code that would be called a fire code and administered by persons that would have expertise in this area, perhaps fire protection engineers. Would this not make for a better system rather than saying a building code should be administered by building officials and "check the oil rags by the fire department?"

*J. E. Bihr* (*author's oral closure*)—I think this might make an interesting debate at some national forum. I take the view that building officials and building inspectors are trained in the construction of buildings and that the fire safety features of buildings are an integral part of building construction. I also feel that they can work cooperatively with the fire services to achieve mutual objectives.

*G. M. Lanier*[4] (*written discussion*)—In the real world, we actually need one set of codes or standards that can be applied to existing situations and a separate set that deals with proposed or new construction. This would be a great asset to enforcement agencies in dealing with the fire problems. We must realize that existing buildings will never be brought into compliance with new criteria. What is a realistic level of risk that can be established where existing conditions are involved?

The response made to Mr. Hammerman's comments and question needs to be reviewed. In my city, the building official and myself have a good working relationship. I feel lucky because it is a fact that, in most areas, building officials and fire officials do not communicate and jointly review plans. In some cases, the building official gives fire protection features proper attention; however, in most cases, fire protection features are loosely

---

[3] Chief fire protection engineer, Office of the Maryland Fire Marshal, Baltimore, MD. 21201.
[4] Fire marshal, Rome Fire Department, Rome Ga. 30161.

applied by building officials. This is primarily due to an underestimation of fire growth, fire spread rates, smoke production, smoke spread, etc. There is also a lack of understanding of how people react to smoke and fire and of the fire extinguishment problems facing the fire department. It is vitally important, in my judgment, that we must get the building officials and fire officials together in the plan review process. Unless we get together, the public will go lacking.

It is suggested that building codes and fire codes contain a requirement for the enforcing authorities to jointly review plans and act together on other matters. For example, the building code would require the building official to consult with the fire official on fire protection matters, and the fire code, on the other hand, would require the fire official to consult the building official on various matters. Do you feel this is reasonable and realistic?

*J. E. Bihr (author's written closure)*—I heartily concur that a positive working relationship between building and fire officials in a community is of the greatest importance in realistically interpreting and applying building and fire codes. Joint plan review is an excellent way to accomplish this objective. Certainly, there should be an intimate working relationship in both professions to assure that a proposed structure is adequately designed in terms of fire defenses. I concur with your view that it would be inappropriate to strictly departmentalize the roles of the fire and building official.

There is no doubt that existing buildings play the primary role in fire problems to date. These older buildings simply do not have the benefit of modern fire protection techniques. Communities should actively concern themselves with these obvious deficiencies, since these are the buildings that lead the list in fire statistics. Solutions to the fire problems in these buildings must be unique; however, there are tools available.

*R. Friedman[5] (written discussion)*—In your opinion, if a new building is to be fully sprinklered, should present building codes be relaxed in some degree with regard to compartmentation or other features? If so, how? If not, why not?

*J. E. Bihr (author's written discussion)*—Codes already recognize the usefulness of sprinklers in a variety of situations. It is not correct, in my view, to assume requirements are relaxed but rather that the usefulness of a sprinklering system has been recognized as either an alternative or primary means of mitigating hazards. The code approach in this area is quite detailed and encompasses such things as increased building areas, windowless buildings, greater exit travel and so on. There are many building designs in which sprinkler installations were not previously considered due to cost deterrents. The time has now come for these buildings to have the additional protection afforded by the sprinkler systems. In my view, major structures,

[5] Scientific director, Factory Mutual Research Corp., Norwood, Mass. 02062.

fully sprinklered, do not warrant significant reductions in key fire resistant features because redundancy is necessary should sprinklers become inoperable through damage, abuse, or other causes. It is simply not prudent to ignore these possibilities.

*A. Peppin*[6] (*written discussion*)—In your presentation, you raised the issue that, although there is a need to evaluate interior finishes for flame spread, ASTM Method E 84 will not necessarily predict the ultimate performance of such materials. Would you care to expand on this? Do you see the need for improved tests that would, in fact, predict performance? If so, what type of tests would you envisage? (for example, full-scale tests?)

*J. E. Bihr* (*author's written closure*)—ASTM Method E 84 presents a basis of comparison of the surface flammability of materials under standard conditions of tests. The test method is not designed to predict the performance of materials in various room combinations acting together with different finishes or with different distributions nor does it examine synergistic effects. Full-scale room burn-out tests are the only realistic method, in my opinion, to evaluate the contribution of specific finish deployment in a room fire, and even this may be misleading due to ventilation, ratio of noncombustible contents to total combustion, etc. ASTM Method E 84 is a useful tool in characterizing the materials that are tested in the fuller scale tests, but ASTM Method E 84 does not provide a basis for judging total fire performance in a building environment. Codes recognize this and prescribe additional limits on finish beyond ASTM Method E 84 results.

[6] Commercial development manager, Monsanto Company, St. Louis, Mo. 63166.

*R. E. Stevens*[1]

# A Place for Voluntary Standards

**REFERENCE:** Stevens, R. E., "**A Place for Voluntary Standards,**" *Fire Standards and Safety, ASTM STP 614,* A. F. Robertson, Ed., American Society for Testing and Materials, 1977, pp. 222–230.

**ABSTRACT:** Voluntary standards are defined as standards used voluntarily. Of the more than 23 000 voluntary engineering standards in use today, most are commercial standards. The preparation of these standards represents the contributions of thousands of experts who write "consensus" standards that in some way affect every product, service, process, or environment. The future of voluntary standards will be assured if those standards are adequate and their use satisfactory.

**KEY WORDS:** fires, standards, engineering standards

For more than three quarters of a century, private organizations like the American Society for Testing and Materials (ASTM), the National Fire Protection Association (NFPA), and many others have developed and published standards. So great was the proliferation of standards that, as early as 1918, people realized a need to (*a*) coordinate these efforts to avoid duplication and to generally put some semblance of order into the burgeoning development that was taking place and (*b*) to provide some acceptable and easily identifiable mark of credibility to standards developed by trade associations and other groups where such might otherwise be lacking. That realization resulted in what we know now as the American National Standards Institute (ANSI).

In 1972, the National Bureau of Standards (NBS) listed 23 300 voluntary engineering standards.[2] If one considers that most of these standards are written by committees, it is apparent that literally thousands of people are involved in standards making. Consider also that most of these persons serve voluntarily; that is, they do not charge the organization sponsoring

---

[1]Assistant vice president, Standards, National Fire Protection Association, Boston, Mass. 02210.
[2]National Bureau of Standards Special Publication 329, an index of U.S. Voluntary Engineering Standards. In 1945, the National Directory of Commodity Specifications listed 35 000 specifications, but that publication has not been updated.

the committee on which they serve for their services. Therefore, the beneficiaries or users of the committee product are largely receiving a free service. While the beneficiaries may be on the receiving end, the employer of the committee member (or the member if he is self-employed) is on the giving end. The travel expense, the committee member's salary, and the cost to the employer for time while the member is away, multiplied by the number of people who serve on technical committees, without doubt represents many millions of dollars annually. It is doubtful that anyone has attempted to estimate the amount of money that is spent annually on standards writing by the private sector.

Not measureable in monetary terms is the vast resource of knowledge and skills that these committee members provide; it is staggering to consider and probably unmatched by any other voluntary effort.

This voluntary contribution, interestingly, is not simply from profit-making organizations. It is also from nonprofit organizations and associations, Federal, State, county and municipal agencies, educational institutions, and the like. Certainly, there is some motivation that prompts these people to participate, and undoubtedly that motivation varies, depending upon the interest of the employer of the committee member. In some cases, that motivation is simply an interest in contributing.

Voluntary standards cover practically every product, service, and practice. They affect commerce, industry, and public safety, and thus, ultimately, every inhabitant of the United States. Standardization alone has made a tremendous economic impact, and, without it, there would be chaos. It is not possible to estimate the injuries and loss of life and property that the voluntary standards have been responsible for preventing.

It seems axiomatic that the voluntary standards have made a tremendous contribution to the well being of this nation.

## Voluntary Standards Defined

For the purpose of this discussion, voluntary standards are standards that are used voluntarily. While this is a very definitive statement, the fact is that the situation is not that definitive.

In the private standards world, it is volunteers who prepare the standards. This has prompted some to refer to the standards that volunteers write as voluntary standards. In organizations like the NFPA, volunteers write codes, standards, recommended good practices, manuals, and guides. Of this variety of output, only codes and standards are written in a fashion that makes them suitable for adoption as requirements to be applied as law. The volunteers who prepare these codes and standards are fully aware that they are writing documents that may be adopted by a local, State, or Federal government agency for legal enforcement. These documents then could hardly be defined as voluntary documents. It is a fact, however, that, in areas where those codes and standards, or others like them, have not been promulgated by a government agency, the codes and standards are used voluntarily by industry and business in general, as well as private and public institutions of all kinds. The recommended practices, manuals, and guides are not written for legal application and therefore are voluntary documents.

It is desirable too to place the word "standard" in perspective for the purpose of this discussion. "Standard" is used in its broadest sense as defined in Websters' Third New International Dictionary as "something that is established by authority, custom, or general consent as a model or example to be followed." It is assumed that ANSI considers a standard to be as Webster defined it because ANSI adopts all kinds of documents, some of which could be called textbooks, and labels them American National Standards.

The NFPA, as well as some other standards-making organizations, define standards in a much narrower context (that is, a document containing only mandatory provisions using the word "shall" to indicate requirements). It is this wide differentiation of meaning among standards makers and users that undoubtedly confuses many persons. In the voluntary standards field, however, it is much more common to find documents that fail to meet the NFPA definition but do meet the Webster definition. This fact may be a problem with voluntary standards.

### Use of Voluntary Standards

If one accepts the estimate that there are more than 23 300 voluntary engineering standards published in the United States alone and recognizes that standards are only prepared and published when there is a recognized need, one must postulate that the use of standards must indeed be impressive.

The specification writer, for example, will seek out and specify standards for compliance with the specifications that he writes. The manufacturer, recognizing this fact, actually includes recommended specifications in his literature and indicates the standards which his product meets. This may be true even though his product is not regulated and therefore not required to meet a standard. It provides the specification writer (and any other person interested in the product) with an assurance that the product, in the opinion of the manufacturer, meets a national standard which is important for design purposes and for the purpose of specifying a desired quality level.

Other reasons for manufacturing a product to meet a standard are the recognized need for standardization to allow interchangeability of parts, to enhance modular design, and to promote international trade. Another compelling reason for manufacturing to meet a standard is that, in the event that the product of the manufacturer is involved in or the result of an accident, and liability is incurred, voluntary compliance with the standard, if it is a national standard, would most certainly be weighed by the court in making its decision. Such use of standards can also be translated into the voluntary use of safety standards that are written for adoption and use by regulatory authorities. Frequently, property owners voluntarily comply with such standards because (a) they feel an obligation to provide a safe environment to the people that inhabit buildings for whatever reason and (b) they recognize the necessity to protect their property from, for example, the ravages of fire.

In cases of injury or loss of life, the property owner may be sued by the injured person or the survivors of the person whose life has been lost. In such cases, the owner may have to justify why he did not follow nationally recognized standards to protect the people in his building.

Another compelling reason for voluntarily complying with safety standards pertains to the more favorable insurance rate that a property owner pays.

The aforementioned examples represent only a few of the uses of voluntary standards. Undoubtedly, there are many more uses, and, in fact, there is probably hardly a product, service, process, or environment that is not, in some measure, subject to a standard of some sort. It seems incredible that anyone could ask the question: Is there a place for voluntary standards?

**Current Voluntary Standard Situation**

Today, perhaps, as never before, voluntary standards are subject to scrutiny by all sorts of persons. The public has been made to believe by certain consumer activists and feather-bedding bureaucrats that some voluntary standards are not adequate and that the application of them produces inferior, sometimes dangerous results. One activist has stated that the responsibility for developing standards which eventually become embodied in public statutes should rest with a governmental body which is accountable and open to the public and subject to conventional legal controls. It has been charged that committees that write voluntary standards are dominated by those who are subject to the standards and that there is little or no representation on committees of the consumer interest. An interesting result of this latter charge is the conglomerate of organizations that now purport to represent the consumer. Some of them are trade associations. Organizations that have been publishing voluntary standards for years, some of them for more than three quarters of a century, are now faced with allegations of not providing adequate "due process" in their standards-making systems, and, therefore, the results are not "consensus" standards. Demands are made that standards be accompanied by cost-benefit and cost-effectiveness studies, and environmental impact statements and that research studies or statistical data provide the basis for each requirement. Standads based on engineering judgement are not adequate in the opinion of some persons.

Added to these pressures on standards-making organizations are the events taking place at the Federal Government level. Two such events that have occurred within the last few years are the Occupational Safety and Health Act (OSHA) and the Consumer Product Safety Act (CPSA). The impact of OSHA has already been felt, and, fortunately for the voluntary standards world, the act permitted the use of "consensus" standards initially. Even that brought forth complaints from some sectors. For example, one large segment of industry that had been participating in the committee activity which produced the consensus standards that were included in the OSHA regulations and which had agreed to those standards was heard to remark, "Yes, we voted for the standards, but we never thought we would have to comply with them." What will happen now that the original OSHA regulations are being revised is difficult to predict, but it is likely that the regulations and the standards (which are also being revised) will drift farther and farther apart as time passes.

The impact of the CPSA has really not been felt yet, but it seems unlikely that the Consumer Product Safety Commission (CPSC) will adopt voluntary standards for promulgation. The act is so broad in its scope that, in years to come, the commission could promulgate many standards covering all kinds of products.

There has been other legislation passed that has affected voluntary standards on a much smaller scale than OSHA or CPSC. The distressing part of this legislation is that it has not recognized the usefulness of the voluntary standards effort of this nation. Most of it has been so written that any standard resulting from the legislation preempts all local and state standards. This means that the voluntary standards sector cannot compete in this standards market. The legislation has therefore preempted states rights and eliminated free enterprise.

How are the standards-making organizations responding to these pressures? Standards-making systems are being revised to assure due process: the classification by interest of committee members is being carefully scrutinized; more consumers are being sought for committee membership; and more public review of proposed standards is inherent in the revised systems. Coupled with these changes are provisions for the recognized need to publish standards more expeditiously. Every effort is being made to provide input to the proposed revisions to the OSHA regulations so that they will be as compatible as possible with current editions of voluntary standards.

At least one standards-making organization has been an "offerer" to CPSC for the writing of a standard for that commission and has adjusted its normal standards-making procedures to accommodate the extremely short time constraints for developing a standard.

As for the committee's writing standards, they are receiving more statistical data than ever before, and they are promoting research and using research results.  They are also writing performance-based standards wherever possible, and they recognize that, in the near future all standards must be revised to reflect the use of Système Internationale units.

It is truly remarkable that committees respond so willingly to the demands made on them by the organizations that publish the standards they write. This response clearly is a reflection of the ability of the committee members to recognize the need to respond to current needs and a reflection of their dedication to the voluntary standard efforts.

**The Future of Voluntary Standards**

The future of voluntary standards depends largely on (*a*) the need for voluntary standards as realized by their users and (*b*) the continuation of the support for the voluntary standards-making organizations as reflected, for example, by participation in committee work.

To date, there has been no evidence that the need for voluntary standards has diminished.  On the contrary, new voluntary standards are being developed, and the development was only instigated after demonstrated need. New materials, new processes, new practices all create a need for new standards.  Unless technological development and growth come to a complete standstill (which state would affect much more than voluntary standards), there will always be a need and a future for voluntary standards.

It will probably be noted that the preceding sentence did not include the phrase "unless the Federal Government assumes all the standards development functions for the nation."  Of course, this is a possibility, but, for many reasons, some of which are given in this paper, the possibility of the Federal Government assuming that role appears extremely remote.  Obviously, it behooves the private sector to avoid actions which would provide the advocates of the Federal Government taking such a role with sound reasons for their advocacy.

Nor has there been any significant diminution to date of support of organizations that publish voluntary standards.  There are, however, indicators of how sincere that support is when the Federal Government decides to promulgate a standard in an area that has traditionally been served by voluntary standards. Experience has shown that, in some of those cases, support from industry, trade associations, designers, and others quickly sways to the Federal Government, apparently because (*a*) Federal law, which preempts all state and local law, appears to be a simple way to eliminate the possibility of there being local regulations that are not uniform, and (*b*) strong private organizations can instigate pressure for change from politicians if it appears that the Federal regulations are unacceptable to those organizations.  It is not the purpose in this discussion to dwell on the threat that such activities by the Federal Government and the private sector alike make to states rights and the free enterprise system.

The question, as it pertains to the future of voluntary standards, is why it was felt necessary that there be Federal legislation. Why wasn't the voluntary standard adequate and its voluntary use satisfactory? These are the questions that must be asked of each voluntary standard: Is it adequate (that is, does it fulfill the needs of all), and is its use satisfactory (that is, is it, in fact, being used universally and voluntarily complied with)?

## Summary

There is indeed a place for voluntary standards, now and in the future. The current use of voluntary standards is impressive and growing. But voluntary standards are also under more public scrutiny than ever before, and those organizations that develop, process, and publish standards must adjust to meet societal needs now.

The future of voluntary standards depends upon the need for them and the support received by the organizations that develop, process, and publish standards. The Federal Government will not take over all the standards development for the nation; it cannot afford to, and the public cannot afford to let it try. The questions to be asked and dealt with now are: Is each voluntary standard adequate, and is its use satisfactory? If these questions can be answered affirmatively, there will be a place for voluntary standards in the future.

# DISCUSSION

*I. N. Einhorn*[1] *(oral discussion)*—I agree with the use of both voluntary and mandatory standards where necessary, but I'd like to hear your comments pertaining to one area that I have been concerned with, and that is assessment of the value of these standards. Do you believe that the role of the voluntary standards organization should include an assessment of the use of these standards? For instance, if one requires detectors or fire suppression equipment or even acceptibility standards for materials, shouldn't there be some sort of evaluation to see whether the standards accomplish their objectives, or, if further modification of the standard becomes necessary, is there a means by which the standards organization will realize this need?

*R. E. Stevens* (*author's oral closure*)—There should indeed, and the feedback mechanism is a vital part of the standards-making system which probably has not been treated as carefully and as well as it should be. However, if a governmental agency adopts a standard and makes it a law, prom-

[1] Professor, Flammability Research Center, University of Utah, Salt Lake City, Utah 84108.

ulgates it, it's very difficult, if not impossible, for the standards-making organization to monitor the activity in the use of that standard, whether it is, in fact, being applied properly, interpreted properly, and so forth. In our organization, we provide an interpretation service, as most organizations do. We have field people who go out and talk to the enforcing people and try to educate the users of the standard. The feedback mechanism is vitally important, and it requires input from all of us, whether or not we are on the committee that writes the document. We should get our thoughts in, get them kicked around, and get them into the standard if they are that important.

*K. P. Reynolds*[2] *(oral discussion)*—I always hear the argument in our own profession that business and industry dominate the fire codes and the standards. This might sound like heresy, but I always think that we, in the fire service, are great Monday-morning quarterbacks. After everything has been done, then we complain about it. If you'll notice, there is very little participation from the fire service in the code and standards development process. In fact, it is appalling how small a fraction of the fire protection enforcement community belong to the NFPA. I wonder if you could just speak briefly on how the NFPA does want enforcement participation in the development of codes and standards?

*R. E. Stevens (author's oral closure)*—Of course, the NFPA is vitally interested in participation by the fire marshals, the fire service, the building officials, the electrical inspectors, and everyone that uses the code either for local purposes or even, on the other end, those that are paying the bill, so to speak. This is a severe problem because of the economic situation within states and cities. It was hard enough a few years ago to get travel money to come to meetings. It's now doubly hard. This is one of the reasons why we changed our standards-making systems to provide a mechanism whereby eveyone would have a formal input. The distribution of our committee reports for each meeting, with the new system, went from 6000 to 12 000 copies. Recognizing that many people cannot participate in committees because of travel money or constraints, they do have an input, and that input has to be considered by the committee, and the committee has to document its action, and we have to publish the results. Yes, we are vitally interested in participation of the fire service.

*W. A. Dunlap*[3] *(oral discussion)*—You mentioned the attempts to get more consumer participation in the standards process, and I appreciate that, but I'd like your comments on the feasibility of that in these very technical kind of matters. When I go to my doctor, and I don't like what he says, I may move to another doctor, but I'm not going to another layman.

*R. E. Stevens (author's oral closure)*—This is indeed difficult. We've had

[2] County fire marshal, Albemarle County, Charlottesville, Va. 22901.
[3] Codes and standards specialist, The Dow Chemical Company, Midland, Mich. 48640.

some rather interesting experiences beating the bushes for consumers on our committees. It is one of the things we've got to do. We have set up what we call a consumer sounding board under the auspices of ANSI. A sounding board is a mechanism by which one of our committees can present its standard or document to a group of consumers. They are housewives or anybody who is simply a consumer. They study the document and come back with critiques of it and input for the committee. But, on the other hand, we have had some real bonafide consumer types on committees. Quite honestly, we have been amazed at the input these people have made, even on some rather technical subjects. They do make a great contribution. The other thing is they don't have travel money either. It is very hard to get a housewife to come to a committee meeting.

*Walter Thomas, Jr.,*[1] *and C. L. Willis*[1]

# National Mandatory Fire Safety Standards*

**REFERENCE:** Thomas, Walter, Jr., and Willis, C. L., "**National Mandatory Fire Safety Standards,**" *Fire Standards and Safety, ASTM STP 614,* A. F. Robertson, Ed., American Society for Testing and Materials, 1977, pp. 231–252.

**ABSTRACT:** The Consumer Product Safety Commission has broad regulatory authority over thousands of consumer products. Mandatory safety standards for matchbooks, aluminum wire systems, and television receivers are being developed or planned under the Consumer Product Safety Act, and two other standards will have significant fire safety aspects. Regulation under the Flammable Fabrics Act has resulted in mandatory standards for children's sleepwear, clothing, carpets and rugs, and mattresses.

**KEY WORDS:** fires, clothing, bedding equipment, burning rate, electric wire, fire detection systems, fire extinguishers, fire hazards, fire resistant textiles, flammability, floor coverings, regulations, heating equipment, television receivers

The Consumer Product Safety Commission (CPSC) has broad regulatory authority over thousands of consumer products. This paper will present an introduction to the fire safety aspects of existing regulations and will discuss rule making in progress within the framework of the legislation administered by the Commission. Also to be discussed are some of the powers of the Commission and the implications of that authority for the community of persons interested in fire safety.

[1] Program manager for flammable products and safety engineer, respectively, Bureau of Engineering Sciences, U.S. Consumer Product Safety Commission, Washington, D.C. 20207.
*This paper was written by Walter Thomas, Jr., and C. L. Willis of the Bureau of Engineering Sciences of the Consumer Product Safety Commission. Inasmuch as it was written by Mr. Willis and Mr. Thomas in their official capacities, it is in the public domain and may be freely copied or reprinted. Opinions expressed in this paper are those of the authors. They do not necessarily represent the official position of the Consumer Product Safety Commission.

### The Consumer Product Safety Commission

*Background*

The Consumer Product Safety Commission resulted in part from the work of the National Commission on Product Safety, created in 1968. The National Commission estimated in 1970 that 30 000 consumers are killed and 20 million injured in accidents in and around the home, many of them product related, resulting in an annual cost to the nation perhaps exceeding $5.5 billion [*1*].[2] In fiscal year 1975, hospital emergency room treatment was given for an estimated 6 936 520 product-related injuries [*2*].

In October 1972, Congress passed the Consumer Product Safety Act (CPSA)[3] which established the CPSC and gave the new Commission authority to set mandatory nationwide safety standards for consumer products.

In addition to its responsibility for administering the CPSA, the Commission was given by transfer the administration of the Federal Hazardous Substances Act,[4] the Flammable Fabrics Act,[5] the Poison Prevention Packaging Act of 1970,[6] and the Refrigerator Safety Act of 1956.[7]

*Data Sources*

In order to help identify the nature, frequency, and severity of product-associated injuries, the Commission has a tool called the National Electronic Injury Surveillance System (NEISS: pronounced "nice"). The NEISS is a computer system linked to the emergency rooms of 119 hospitals statistically selected throughout the nation. The product-associated injuries treated in these emergency rooms are coded in a standard epidemiologic manner which includes the product or product grouping.

In addition to the NEISS data, the Commission receives, from each of the states, death certificates for all product-associated deaths. These supplement the data on injuries which appear in the NEISS emergency rooms.

Fire-related injury data collected in hospital emergency rooms may not accurately reflect occurrences because very serious burns may be treated in special hospital burn units. Victims who die may never be taken to the

---

[2] The italic numbers in brackets refer to the list of references appended to this paper.

[3] Public Law 92-573; 86 Statute 1207, and those that follow; 15 USC 2051, and those that follow.

[4] 15 USC 1261, and those that follow, as amended.

[5] 15 USC 1191, and those that follow, as amended.

[6] 15 USC 1471, and those that follow.

[7] 15 USC, 1211, and those that follow.

emergency room. Some burn incidents may be reported as accidents that do not indicate the occurrence of a fire. The latter may be reported under other product codes as "unknown origin."

One such additional source of data is the Flammable Fabrics Accident Case and Testing System (FFACTS) that was developed at the National Bureau of Standards (NBS) to process and analyze data from reports of individual fires involving fabric products. Between 1967 and December, 1974, approximately 3000 cases from the Food and Drug Administration, the CPSC, and state and local fire marshals were coded into the FFACTS and supplemented by laboratory evaluations. Another source of data will be the National Fire Data System maintained by the National Fire Prevention and Control Administration. Other sources have been the various fire loss reports compiled by the National Fire Protection Association[8] and a survey conducted by the Bureau of the Census funded jointly by the CPSC and the NBS.[9]

## Consumer Product Safety Act

### Overview

Under the CPSA, the Commission can act to prevent or reduce an unreasonable risk of injury associated with a consumer product.[10] The Commission's regulatory authority over consumer products may take various forms, depending upon the nature of the risk. Mandatory nationwide safety standards, which preempt local and state laws, may be promulgated. Consumer products which present unreasonable risks of injury and for which no feasible safety standard would adequately protect the public may be banned. A consumer product in the possession of a manufacturer, distributor, or retailer may be seized if it presents imminent and unreasonable risk of death, serious illness, or severe personal injury. Notification to the public and repair of defects may be ordered if it is found, after a hearing, that a consumer product creates a substantial product

---

[8] See any issue of *Fire Journal,* published bimonthly by the National Fire Protection Association (NFPA), Boston, Mass.

[9] This survey of over 33 000 U.S. households was conducted during the week of 15 April 1974. Those surveyed were asked to recall all types of fire incidents and fire-related injuries or deaths or both since 1 April 1973. A draft of the final report has been printed (Ref 3).

[10] Section 3(a) (1) of the CPSA defines "consumer product" to mean "... any article, or component part thereof, produced or distributed (i) for sale to a consumer for use in or around a permanent or temporary household or residence, a school, in recreation, or otherwise, or (ii) for the personal use, consumption, or enjoyment of a consumer in or around a permanent or temporary household or residence, a school, in recreation, or otherwise ..." Nine specific exclusions from this definition are listed. Among these exclusions are products and substances regulated by other federal agencies, tobacco and tobacco products, and "... any article which is not customarily produced or distributed for sale to, or use or consumption by, or enjoyment of, a consumer ...."

hazard. Civil and criminal penalties may be imposed for violation of a commission rule.

A unique feature of the CPSA is the offeror process by which the Commission publicly solicits offers from all interested organizations and individuals to develop recommended standards or to submit existing standards for consideration by the Commission. This solicitation, published in the *Federal Register,* identifies one or more product-associated unreasonable risks of injury which are to be considered by the offeror during the development of the recommended standard. Once the offeror recommends a standard to the Commission, the Commission must either publish a proposed regulation for the consumer product or withdraw the proceeding for the development of the regulation.

The CPSC is not limited to regulatory action. It also encourages voluntary standards making to reduce unreasonable risks of injury associated with consumer products. The Commission also has an active Bureau of Information and Education which, among other activities, seeks to educate the public about product-related hazards.

## Present Activities in the Fire Safety Field

*Matchbooks*—A standard which is being developed almost exclusively for fire safety is one for matchbooks. The *Federal Register* notice of proceeding listed seven types of injury which were to be considered by the offeror.[11] Injuries to children were specifically included. Some of these injuries are associated purely with the construction of the match or matchbook, such as burns or other injuries resulting from fragmenting matchheads, delayed ignition, sparking, and unexpected ignition. Other injuries are of a type which involve the additional element of human factors, such as the use of matchbooks by impaired persons and children, ignition of the remaining matches in the book when a single one is struck, and a failure to completely extinguish a match that has an extended burn time or demonstrates afterglow.

The American Society for Testing and Materials was the offeror on the matchbook standard. The CPSC itself has developed an additional requirement which, if adopted, would require matchbooks to be child resistant. Under this version, certain design options for the cover would be specifically permitted. The cover could be of a latching type which requires simultaneous motions, sequential motions, or a minimum force for opening. Alternatively, it could be of a nonlatching type if the Commission approves the specific cover design based on its child resistance. A proposed standard may be published shortly in the *Federal Register.*

---

[11] "Bookmatches: Proposed Safety Standards," *Federal Register,* Vol. 39, Consumer Product Safety Commission, 4 Sept. 1974, p. 32050.

*Aluminum Wiring Systems*—Another product for which standard development has been commenced is aluminum wire for residential electrical branch circuits, as announced by the Commission's notice of proceeding on 4 Nov. 1975.[12] A full explanation of the theories which have been suggested to describe the exact manner in which connections made with aluminum wire can overheat and start fires is beyond the scope of this paper.[13]

The CPSC has collected numerous reports of aluminum wire incidents involving electrical malfunctions in single family dwellings, mobile homes, and multifamily dwellings between 1967 and 1975. In these reports, damage ranged from failure of an electrical component to fires with extensive structural damage. Twelve deaths were reported [5].

When aluminum is exposed to the air, an oxide coating of between 30 and 100 Å develops in seconds. This coating is hard, brittle, and has a high electrical resistance. Therefore, to make an electrical connection to aluminum wire, the oxide must be broken and the connection made to the aluminum itself.

The CPSC technical staff believes that the potential fire hazard at wire-binding screw terminals of outlet devices develops when the connection moves,[14] the oxide film forms on the newly exposed surfaces, and the resistance of the connection increases. The combination of aluminum's relatively high coefficient of thermal expansion, and the compressive stress relaxation of the wire have contributed to the movement at connections. With continued motion comes overheating produced by the increasing electrical resistance. The insulating material of the wire and receptacle or switch may then overheat, causing a fire.

Because convenience outlets typically are wired in a series-parallel combination, current may flow through portions of one or more unused outlets to feed another which is being used. For this reason, fires may originate even in unused receptacles [5].

The products to be addressed by the standard are the aluminum conductor in Nos. 10 and 12 Awg sizes and termination devices such as wall switches, convenience outlets, wire splicing devices, and circuit breakers. Copper clad aluminum conductor where the copper coating is 10 percent

[12] "Aluminum Wire Connections: Proceeding for Development of Consumer Product Safety Standard," *Federal Register,* Vol. 40, Consumer Product Safety Commission, 4 Nov. 1975, p. 51218. Among other items, this invitation for offerors presents the Commission's preliminary determination of unreasonable risk of injury associated with the use of aluminum wire, a description of the nature of the risk of injury, and discussion of relevant voluntary standards known to the Commission.

[13] For a compendium of information describing experimental research at the NBS on terminating aluminum wires and also containing information on failure mechanisms, connector cycling tests, and a bibliography, see Ref *4.*

[14] Some causes of relative motion between wire and termination device have been identified as mechanical disturbances, metallic creep in the wire, thermal cycling (caused by the environment or by current flow), and transient electrical surges.

or greater of the conducting cross sectional area has been excluded from the scope of the standard.

In December 1975, the Commission rejected the one offer that it received to develop the standard and directed the staff to develop a plan for writing the standard within CPSC. In addition to the development of the mandatory standard, the Commission in August 1975 authorized the staff to initiate proceedings before an administrative law judge to determine if "older technology aluminum wire systems" [15] installed in 2 million homes presents a substantial product hazard and, if so, to determine what remedial action should be taken by the parties to the proceeding.

*Television Receivers*—Underwriters' Laboratories has been selected as the offeror to develop a safety standard for television receivers. Among the risks of injury to be considered are those from fire.

Unfortunately, in cases of television fires, it is often difficult or impossible to determine the exact cause of the incident. As far back as the late 1960's, characteristics of certain televisions were suspect. Among the influencing factors were the higher operating temperatures and voltages of color televisions and the trend toward the use of polymers. Color televisions operate at electric potentials up to 30 kV, black and white up to 16 kV. Operating temperatures of color televisions are in the 43 to 46 °C range (110 to 115 °F). Certain insulation materials were thought to be susceptible to thermal breakdown or smoldering. Use of combustible materials both in the circuitry and in the enclosures were blamed for contributing to the buildup of fire and noxious products of combustion [6–8].

In order to understand more thoroughly the factors which contribute to the ignition itself and to the severity of the fire (as well as hazards not related to fire), the Commission subpoenaed from television manufacturers reports of accidents and safety-related incidents. Over 10 000 of these reports had been received by the manufacturers over a period from 1970 to mid-1974.

An analysis of 7620 of these incidents showed that 83 percent were fires or fire related. Fifty two of the fires involved fatalities. A higher incidence of fires occurred in damp climates than dry. For those reports in which the set was identified, both as to year of manufacture and as to being black and white or color, the incidence of fires for each year of manufacture was always significantly greater in the color sets and averaged two times that of the black and white sets. The reports also indi-

[15] "Older technology aluminum wire systems" is a shorthand method of referencing aluminum wire in Nos. 10 and 12 Awg generally manufactured prior to 1971, but possibly installed after 1971, and devices manufactured for use with the wiring or that reasonably could be expected to have been used with the wiring. Some attempts to find solutions to the hazards posed by aluminum wire—such as indium plating, low-creep aluminum wire, the CO/ALR termination, and the crimp connection—are not considered "older technology."

cate that a fire is over six times as probable in a set equipped with an "instant-on" feature by which the filaments, including that of the picture tube, are kept energized even when the set is turned off. This figure is based on 861 fires in which the presence or absence of "instant on" was known. In 3103 fires, the origin of the fire within the set was identified; of these, 36.4 percent cited the high voltage section as the origin. The incidence of picture tube implosions during fires is not known, but an implosion was indicated as having occurred in 306 of the reported fires. These reports mentioned implosions as a factor in the spread of fire through the scattering of hot or burning particles or drops of burning, molten plastic. Explosions were said to have been caused when air and unburned gases were suddenly mixed by an implosion [7].

In its development of the proposed safety standard, the offeror was asked by the Commission[16] to consider the possibilities of fire resulting from arcing or overheating of components in conjunction with flammable materials used in the construction of the set, especially the enclosure. Other risks to be addressed are capacitors which remain charged after the set is turned off, the failure of external components such as power supply cords and remote control cords, and picture tube implosions during a fire.

In addition, the offeror will address risks of electric shock from accessible portions of the set. Also, the offeror presently plans to require that a set of installation instructions for achieving proper grounding be given to each purchaser of an outdoor antenna.

*Gas Fired Space Heaters*—Ignition of fabrics, contact burns, carbon monoxide poisoning, and anoxia will be addressed by a planned standard for gas-fired space heaters. Not only can materials such as wearing apparel and draperies be ignited by contact with the flame, but also there are risks from ignition of any gas or vapors that may have accumulated due to a fault in the heater or from other nearby sources. Solutions will be sought for preventing ignition of walls and furnishings from an excess of heat transported through the rear and sides of a space heater. Surface temperature will be regulated to guard against skin contact burns.[17] Assurances of adequate venting and fuel supply also will be included to minimize the danger of carbon monoxide poisoning, anoxia from other causes, and accumulation of combustible mixtures.

Based upon NEISS data, the Commission estimates that space heaters

---

[16] "Hazards from Television Receivers: Proceeding to Develop Safety Standard," *Federal Register*, Vol. 40, Consumer Product Safety Commission, 28 Feb. 1975, p. 8592.

[17] The thermesthesiometer is a device developed by the NBS under CPSC sponsorship. It measures the temperature that would be experienced by a human if contact were made with a heated surface. The measurement, indicated in degrees Celsius on the instrument, is called the "contact temperature." This method of measuring surface temperature has been adopted in ANSI Z21.1 "Household Cooking Gas Appliances" and is being considered for use in UL 858 "Household Electric Ranges."

were involved in 5300 injuries treated in hospital emergency rooms during 1973. Of the emergency room cases in which the type of heater was reported, the greatest number (34 percent) were fueled with gas. The second most prevalent type given was electric powered (12 percent). In 47 percent of the injuries, the type was not given. Other death and injury information available to the Commission from in-depth investigations and death certificates also indicates a high rate of injurious or fatal accidents involving gas [9].

The Commission has determined that only gas space heaters present an unreasonable risk of injury. This determination was based on the just mentioned statistics and on the fact that gas space heaters are associated with serious hazards that are not present in electric space heaters, such as carbon monoxide poisoning, anoxia, and explosion.[18]

Of all the NEISS space heater related injuries, 69 percent were burns. Over half of these were burns to the extremities. This high proportion of burn injuries becomes even more significant when one recalls that burn injuries in general may be underreported in the NEISS.

The Commission has conducted in-depth investigations of over 200 injuries that were associated with gas-fired space heaters [9]. Although distribution of these injury types in the in-depth investigations is not necessarily representative of their distribution within the NEISS, they provide a more detailed indication of types of accidents that are occurring. An analysis of 218 of these investigations showed that 135 were burns that resulted from contact with a surface of the heater. Thirty eight of the injuries involved clothing ignition. Another 16 were carbon monoxide poisonings. The remaining 29 are divided between explosions, ignition of vapors from flammable liquids, nonburn injuries from contact with the heater (for example, lacerations), ignition by the heater of something other than clothing, and miscellaneous.

The American National Standards Institute (ANSI) and Underwriters' Laboratories (UL) have promulgated standards which govern, among other things, aspects of safety associated with gas fired heaters.[19] Following a decision by the Commissioners to do so, an invitation for offerors to develop a standard or submit existing standards is planned for publication in the *Federal Register.* Any such notice will contain a discussion of what the Commission believes are the merits and inadequacies of these existing standards. In the meantime, the staff has been following en-

---

[18] "Space Heaters: Partial Denial of Petition," *Federal Register,* Vol. 40, Consumer Product Safety Commission, 5 Sept. 1975, p. 41172.

[19] These include the National Fuel Gas Code (ANSI Z223.1–1974, NFPA No. 54); "Standard for Gas-Heating Appliances for Mobile Homes and Travel Trailers, UL 307(b); "American National Standard for Gas-Fired Room Heaters, Vol. I, Vented Room Heaters, ANSI Z21.11.1–1974; "American National Standard for Gas-Fired Gravity and Fan Type Direct Vent Wall Furnaces, ANSI Z21.44–1973; Amerian National Standard for Gas Fired Gravity and Fan Type Vented Wall Furnaces, ANSI Z21.49–1975.

couraging changes in four standards of the American National Standards Committee on Performance and Installation of Gas Burning Appliances and Related Accessories (Committee Z21).

*Other*—The Commission under the CPSA may regulate fire-related risks of injury other than sources of accidental ignition. For example, there are plans to initiate the offerer process for the development of a fire safety standard for tents.

## Research

Where necessary to an understanding of the risks or injury modes, the Commission conducts research directed toward improving the safety of the products involved. Two such programs are being conducted for the Commission by the National Bureau of Standards.

Because consumers continually are injured in accidents involving flammable fabrics and related materials, the Commission has a broad-based research program at NBS consisting of basic flammability and hazards, analysis of injury data, and development of test concepts for general wearing apparel and textile interior furnishings not covered by existing standards. The work involves research into the mechanisms of fabric combustion (including smoldering) and flame inhibition. Also covered are the characteristics of full-scale fires involving textile interior furnishings and correlation of small- and large-scale tests.

The CPSC has joined with the Society of the Plastics Industry, the Department of the Navy, and the Veterans Administration in a program at NBS to address the difficult problem area of generic fire hazards associated with plastic items generally and with items used to furnish and decorate the home. Plastic products have been found to be involved in many serious fires, both as fuel and as producers of noxious smoke. The work attempts to identify plastic products involved in fires and characterize the properties that make the products hazardous. Laboratory testing and simulation of fire incidents are employed both on specimens from fire sites and on purchased specimens. Room burn tests of plastic furnishings are conducted in order to assess their role in compartment fires.

Looking to the future, the Commission staff has been considering alternatives to safety devices that do not function properly when needed. The staff has gathered information about the various forms of fire detectors and fire extinguishers, together with case histories of malfunctions. Of special concern are the related problems of failure of a detector to alarm in time during a fire of the sort which should activate a life-protection device or system,[20] and the use of inappropriate detectors for particular situations. For example, studies have shown that detectors

[20]See the following articles in *Fire Journal,* NFPA; Vol. 68, No. 2, March 1974, p. 52; Vol. 68, No. 3, May 1974, p. 65; Vol. 70, No. 1, Jan. 1976, p. 43.

using a metal oxide semiconductor did not alarm under test conditions that activated photoelectric and ionization types within the prescribed time period [*10*].

The engineering staff has followed with interest the development by Bukowski and Bright of performance specifications for single-station smoke detectors. These specifications are a combination of criteria from UL standards and modifications to them based on work done at the NBS.[21] There also are a number of other standards for detectors.[22]

Soda-acid fire extinguishers, no longer manufactured, have been identified as potential hazards based on deaths and injuries sustained by people attempting to discharge the units [*11*]. Pressurization of extinguishers weakened by corrosion and embrittlement can lead to violent rupture of the case or dome. Studies are being conducted to determine the cause of failure and to recommend nondestructive test methods for identifying the dangerous units.[23]

Preliminary investigations continue into other aspects of the residential fire problem. Mostly, this activity has consisted of gathering research results from around the country and gathering fire experience data categorized by product. This work, as well as the work in detectors and extinguishers, is now only a staff function; the Commission has not expressed an intent to take regulatory action in these areas.

The Commission's Bureau of Biomedical Science has been engaged in work designed to assess and investigate the difficult problems that may generically be called the toxicity of products of combustion. One study, conducted for the Commission by Dr. Dressler of the Harvard University Medical School, was designed to develop an acceptable smoke toxicity test and to establish the validity of this test method by determining the relative toxicity of 22 materials commonly found in the home.[24] A controlled-atmosphere smoke apparatus, designed at Harvard, was used to expose 1638 rats to the smoke produced by the flaming or smoldering of varying concentrations of 21 materials. White pine wood was used as a control. The materials burned represented eight major categories of home

[21] The proposed "Standard for Single and Multiple Station Smoke Detectors," UL 217, combines the present requirements appearing in "Standard for Smoke Detectors, Combustion Products Type," UL 167, and those from the proposed fourth edition of "Standard for Photoelectric Type Smoke Detectors," UL 168 (ANSI A131.4–1971).

[22] "Standard for Fire Alarm Devices, Single and Multiple Station, Mechanically Operated Type," UL 539; "Standard for Household Fire Warning System Units," UL 985; "Standard on Automatic Fire Detectors," NFPA No. 72E–1974; "Standard for the Installation, Maintenance and Use of Household Fire Warning Equipment," NFPA No. 74–1974; "Approval Standard Smoke Actuated Detectors for Automatic Fire Alarm Signaling," Factory Mutual Research 3230–3250.

[23] Christ, B. W. and Smith, J. H., "Safety Evaluation of Soda-Acid Fire Extinguishers," NBSIR 75–922, National Bureau of Standards, Washington, D.C. (unpublished).

[24] Dressler, Donald P., MD, *Smoke Toxicity of Common Consumer Products,* Final Report, Harvard Medical School, Department of Surgery, 16 June 1975 (unpublished).

furnishings: wallpaper, carpets, noncarpet floor covering, upholstery, draperies, wallboard, paint, and wood. Harvard reported that the test is reproducible and that differential mortality may be determined from physiological monitoring of the rats and histological examination of the lungs. Follow-up work has been conducted and will be reported shortly.

## Flammable Fabrics Act

### Background

When the Consumer Product Safety Commission was activated on 14 May 1973, the jurisdiction of the Flammable Fabrics Act (FFA), as amended, was transferred to CPSC from the Department of Health, Education, and Welfare, the Department of Commerce, and the Federal Trade Commission. The FFA was established by Congress in 1953 as a result of serious burn injuries incurred in accidents involving flammable apparel. Most notable were the notorious "torch" sweaters and boys' cowboy chaps. The FFA was applicable to all wearing apparel except hats, gloves, footwear, and interlining fabrics.

In 1967, the FFA was amended to extend the coverage of the Act to interior furnishings and other areas of wearing apparel which had been previously excluded. Authority was given to the responsible agency to promulgate flammability standards based on investigations or research and a finding that such standards are needed to protect the public adequately against unreasonable risk of the occurrence of fire leading to death, injury, or significant property damage. The Act states that each standard, regulation, or amendment thereto promulgated shall be reasonable, technologically practicable, and appropriate.

The Commission continues to promulgate standards for flammable fabrics under the FFA.

### Existing Standards

*General Wearing Apparel*—Commercial Standard 191-53, "Flammability of Clothing Textiles," was adopted by Congress in 1953 as the flammability standard for determining compliance of general wearing apparel with the requirement of the FFA. This standard specifies the testing of 2 by 6-in. fabric specimens placed at a 45-deg angle in a test chamber. After mounting the specimens in metal holders, oven drying, and cooling in a desiccator, the specimens are exposed to a small 5/8-in. flame for 1 s, and, if ignited, the rate of flame spread over a 5-in. distance is measured. Fabrics which ignite and burn the 5-in. distance in less than 3.5 s for smooth surface fabrics and less than 4.0 s for raised-fiber surface fabrics are prohibited from sale in commerce.

*Carpets and Rugs*—The Standard for the Surface Flammability of Carpets and Rugs, FF 1-70, and the Standard for the Surface Flammability of Small Carpets and Rugs, FF 2-70, became effective on 16 April 1971 and 29 Dec. 1971, respectively. FF 1-70 is applicable to carpets and rugs which have one dimension greater than 1.83 m (6 ft) and a surface area greater than 2.23 m² (24 ft²), whereas FF 2-70 is applicable to carpets and rugs less than this size. The difference between FF 1-70 and FF 2-70 is that small carpets and rugs that do not comply with the requirements of FF 2-70 can be sold if they are properly labeled.

Both standards (Fig. 1) require that seven of eight specimens, selected

FIG. 1—*Test apparatus for carpet and rug standards FF1-70 and FF2-70.*

as representing the lot, pass the test criterion. Each specimen is oven dried, cooled in a desiccator and placed in a test chamber for testing. A small methenamine tablet is ignited at the center of the specimen. If the resulting char does not extend to within 1 in. of a flattening frame containing an 8-in. diameter hole, the specimen passes the test. Carpets and rugs that contain a fire-retardant treatment must be laundered according to a specified procedure prior to being tested. These treated carpets and rugs must be labeled with the letter "T."

*Children's Sleepwear*—The Standard for the Flammability of Children's Sleepwear, Sizes 0 through 6X, FF 3-71, as amended, and the Standard for the Flammability of Children's Sleepwear, Sizes 7 through 14, FF 5-74, as amended, became effective on 29 July 1972 and 1 May 1975, respectively. These two standards (Fig. 2) involve subjecting sets of

FIG. 2—*Test apparatus for children's sleepwear standards FF3-71 and FF5-74.*

five individual specimens measuring 3½ by 10 in., which have been dried, to an open flame for 3 s. After combustion ceases, the damaged portion of the specimen, the char length, is measured. FF 3-71 requires an additional measurement, the length of time that burning fragments remain on the base of the test chamber. The standards specify that the average char length of the five specimens shall not exceed 7 in., while no individual specimen char length shall be greater than 10 in. For garments and fabrics covered by FF 3-71, no residual flame time shall persist longer than 10 s.

The standards include provisions for testing the durability of flame-resistant characteristics to laundering and labeling to alert consumers to agents that may reduce flame resistance. Also specified in each standard is a sampling plan that requires premarket testing of fabrics and garments before introduction into commerce.

*Mattresses*—Information available at the time the development of the standard for mattresses began showed that the typical hazard associated with accidents involving mattresses was the result of smoking.[25] Typically, the individual would go to sleep with the cigarette still burning; the cigarette then ignited the mattress materials, and the mattress smoldered. The individual, and possibly others in the building, were asphyxiated. Therefore, the cigarette was chosen as the ignition source for the test method in the standard.

The Standard for the Flammability of Mattresses (FF 4-72), as amended, became effective on 22 June 1973, and included a six-month period for labeling of noncomplying mattresses. This warning label amendment was intended to ease the economic impact of the standard on small manufacturers.

FF 4-72, as amended, contains a sampling plan and requires premarket testing by the manufacturer. The standard (Figs. 3a, b) specifies proto-

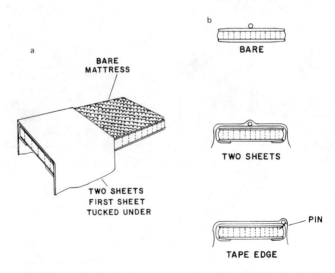

FIG. 3—*Schematic of FF4-72:* (a) *mattress preparation and* (b) *cigarette location.*

[25] See Refs *12* and *13*. Also "Mattresses: Proposed Flammability Standard," *Federal Register,* Vol. 36, No. 18095, Consumer Product Safety Commission, 9 Sept. 1971.

type qualification and production testing at least once every three months for each mattress type. On each required mattress surface, a minimum of nine cigarettes are placed at various specified locations on one half of the bare mattress surface. The test is repeated on the other half of the mattress surface with the cigarettes placed between two cotton sheets. Six surfaces are required for prototype testing and two surfaces for production testing.

*Present Activities*

The staff will present the draft standards discussed later for upholstered furniture, general wearing apparel, and specified wearing apparel for consideration by the Commission as possible approaches for reducing burn injuries associated with those products.

*Upholstered Furniture*—A Finding of Possible Need for a standard regulating upholstered furniture was published by the Department of Commerce in the *Federal Register* on 29 Nov. 1972.

The hazard associated with upholstered furniture is very similar to the scenario described for mattresses. Therefore, the cigarette is being considered for the ignition source in the standard under development for upholstered furniture.

A draft of a recommended standard was submitted to CPSC staff from the NBS in October 1974. The Commission staff have reviewed the draft proposed standard, conducted a preliminary economic impact study, and are formulating a final briefing package for submission of the proposed upholstered furniture flammability (smolder resistance) standard to the Commission for consideration.

The draft standard (Fig. 4) involves testing mock-up representations of furniture items for resistance to cigarette ignition. The mock-up test concept represents a new approach to testing for standards promulgated under the FFA. It is recognized that this concept is compromising assurance that the final product will not ignite from smoldering cigarettes in some configuration and circumstances. However, in an effort to obtain a feasible solution to reducing the primary hazard, the mock-up concept is being considered. Additionally, this concept, if finalized in a standard, will require different techniques for enforcement.

A specified number of cigarettes are placed in different locations on the mock-up. If any of the cigarettes cause the mock-up to ignite or char more than 3 in. away from the cigarette, the furniture design represented by that mock-up is not acceptable. To reduce the furniture manufacturer's testing burden, provisions are included for classifying upholstery fabrics by their performance when exposed to a cigarette. A mock-up can then be qualified for use with an entire class of fabrics with one test. Periodic testing is required to give assurance that the furniture and fabric cigarette ignition characteristics remain acceptable.

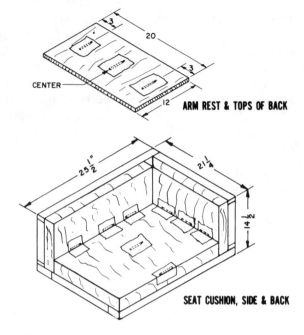

FIG. 4—*Schematic of upholstered furniture test mock-up.*

*General Wearing Apparel*—On 23 Oct. 1968 the Department of Commerce issued a notice of finding of possible need to amend or replace CS 191-53. The commercial standard had been determined to be technically inadequate to protect the public. All 117 garments involved in the burn injury cases of the FFACTS met the existing requirements.[26] Subsequent analysis at NBS of injury data and characteristics of the fabrics involved has also shown that the flame spread rates measured by CS 191-53 do not relate to injury severity [*14-17*].

The CPSC, in light of this need, is supporting research work at NBS to develop a new, more realistic test method for measuring the flammability of wearing apparel. Several promising concepts are being investigated at this time, specifically, ease of ignition, heat transfer, and ease of extinction. A test apparatus, commonly called the "mushroom" (Fig. 5), has been designed to measure the ease of ignition and heat transferred to the body by a burning fabric. The test incorporates a cylindrical fabric specimen attached to a circular sensorized copper cylinder. The specimen configuration produced is much like that of a garment on a body.

The measurement of heat transfer could be used to define self-extin-

---

[26] "Notice of Finding that Flammability Standard or Other Regulation May Be Needed and Institution of Proceedings," *Federal Register,* Vol. 33, No. 15662, Consumer Product Safety Commission, 23 Oct. 1968.

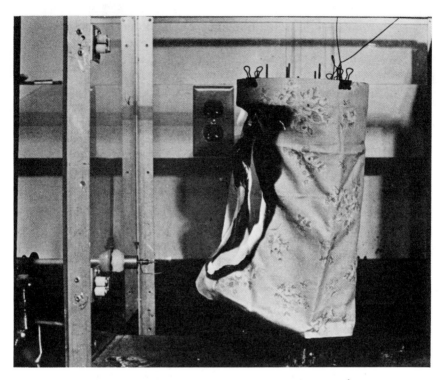

FIG. 5—*New test apparatus for general wearing apparel.*

guishment and to predict the size and depth of burn caused by the burning fabric. It is anticipated that specific heat transfer ratings, ignition, and extinguishment characteristics could be required of fabrics for use in garment types representing compatible risks or potential hazards.

*Specified Wearing Apparel*—Another alternative for addressing the wearing apparel problem now being considered by the Commission staff involves the extension of a modified version of FF 5-74 (Flammability of Children's Sleepwear, Sizes 7 to 14) to other items of apparel. The garment items causing the most severe injuries and cited most frequently in available injury data are nightgowns, robes and housecoats, pajamas, dresses, shirts, and pants [*18*].

For this purpose, the Children's Sleepwear Standard was modified to require only fabric testing, except in cases where excessive amounts of trim are being used on a garment. Test specimens are conditioned at 70°F (21.1°C) and 65 percent relative humidity rather than oven dried. Only 35 launderings are required, rather than 50.

On the basis of this modified standard, Commission staff have conducted an economic impact analysis that estimates the effects of this standard, if promulgated. This report, as well as other background material, will be submitted in a briefing package to the Commission for a decision on whether to publish the standard and to what extent.

*Flammability Test Module*—Another aspect of federal standards is the necessity for their enforcement. In order to provide timely laboratory testing of specimens in the field, the Commission is designing a flammability test module. This apparatus will be 28 in. high, 15 in. deep, and 19 in. wide. It will combine the testing capabilities required for CS 191-53, the carpet and rug, and the sleepwear standards. It is anticipated that the module will be used for screening purposes only since it will not have the capacity for dry cleaning and laundering fabrics or cleaning carpeting.

## Federal Hazardous Substances Act

### Background

The Federal Hazardous Substances Act (FHSA), originally the Federal Hazardous Substances Labeling Act, was enacted 12 July 1960 and later amended by the Child Protection Act of 1966 and the Child Protection and Toy Safety Act of 1969. On 14 May 1973, the CPSA transferred from the Secretary of Health, Education, and Welfare the functions under the FHSA to CPSC. Also transferred were the accompanying regulations promulgated by the Food and Drug Administration.

In part, the Act defines the term "hazardous substances" as: "any substance or mixture of substances which ... (v) is flammable or combustible ... if such substance or mixture of substances may cause substantial personal injury or substantial illness during or as a proximate result of any customary or reasonably foreseeable handling or use, including reasonably foreseeable ingestion by children."[27]

### Existing Regulations

The Act defines "extremely flammable," "flammable," and "combustible" substances (liquids) in terms of flash points obtained with the Tagliabue Open Cup Tester.[28] A procedure for determining flash points by the Tagliabue Open Cup Tester is provided in the regulations for the FHSA. The Act addresses "extremely flammable," "flammable," and "combustible" solids and contents of self-pressurized containers but states that the flammability or combustibility will be established by regulation.

In regulations issued under the FHSA, the Food and Drug Administration published a definition for "extremely flammable" solids and pro-

---

[27] Federal Hazardous Substances Act, Section 2(f).
[28] Federal Hazardous Substances Act, Section 2(1) and 16 CFR 1500.

vided test procedures for "flammable" solids and "extremely flammable" and "flammable" contents of self-pressurized containers.[29] To date, there are no definitions or test procedures for "combustible" solids or "combustible" contents of self-pressurized containers.

These regulations also address the problem of a substance or mixture of substances that "generates pressure through decomposition, heat, or other means" by providing the following criteria.

1. If it explodes when subjected to an electric spark, percussion, or the flame of a burning paraffin candle for 5 s or less.

2. If it expels the closure of its container, or bursts its container, when held at or below 130°F for two days or less.

3. If it erupts from its opened container at a temperature of 130°F or less after having been held in the closed container at 130°F for two days.

4. It it comprises the contents of a self-pressurized container.

## Present Activities

The Commission staff is in the process of developing a program to review all the flammability regulations under the FHSA. Such a program will identify the hazard associated with products or classes of products covered by the regulation and consists of reviewing, modifying, or developing new regulations or all of these, including test methods to more adequately predict the hazards.

Presently, the Commission is funding work at the NBS to review and modify, where appropriate, the flammable solids test method. The test method, as now written, specifies testing of materials by measuring their dimensions, placing the material by means of clamps, ringstand, etc., in a horizontal configuration, and measuring a rate of flame spread. This technique is not appropriate for many nondimensional products such as toys, decorations, Easter grass, Christmas trees, etc. Therefore, there is a need to design a test method or methods that measure the flammability of products in their actual use condition to more accurately predict the hazards. Information on burn incidents is being accumulated to help identify the hazard patterns.

## Summary

The CPSC has broad regulatory authority over thousands of consumer products. Identification of risks is the first step in the regulatory process. Within the fire safety field, this identification can come from injury data, fire reports, tests designed to demonstrate the existence of a hazard, and the work of outside groups involved in fire hazard identification.

[29] 16 CFR 1500.

Regulation under the FFA has resulted in mandatory standards for children's sleepwear, clothing, carpets and rugs, and mattresses. The FHSA contains broad flammability regulation for "extremely flammable," "flammable," and "combustible" substances and solids. Three mandatory fire safety standards under the CPSA are being developed or planned, and two other ones will have fire safety aspects.

Future Commission emphasis under the FFA for clothing and interior furnishings may take the course of generic or general type standards. Under the CPSA and the Federal Hazardous Substances Act, the approach may be either generic or on a product-by-product basis.

## References

[1] "Final Report," National Commission on Product Safety, Washington, D.C., June 1970.
[2] "Annual Report," Consumer Product Safety Commission, Washington, D.C., Oct. 1975.
[3] Analysis of National Household Fire Survey (Draft), Consumer Product Safety Commission, Washington, D.C., June 1975.
[4] "Aluminum Branch Circuit Wiring in Residences Summary Report for the Consumer Product Safety Commission January–September 1974," NBSIR 75-723, National Bureau of Standards, June 1975.
[5] Hazard Analysis of Aluminum Wiring, Consumer Product Safety Commission, Washington, D.C., April 1975.
[6] Sessions, National Commission on Product Safety, Vol. No. 8, Oct.–Dec. 1969, Law-Arts Publishers, New York.
[7] Yereance, R. A., "Report on Analysis of TV Accident Data to Consumer Product Safety Commission," 25 April 1975.
[8] Hazard Analysis of Television Sets, Consumer Product Safety Commission, Washington, D.C., Oct. 1974.
[9] Hazard Analysis of Injuries Related to Space Heaters, Consumer Product Safety Commission, Washington, D.C., Feb. 1975.
[10] Bukowski, R. W. and Bright, R. G., Some Problems Noted in the Use of Taguchi Semiconductor Gas Sensors as Residential Fire/Smoke Detectors, NBSIR 74-591, National Bureau of Standards, Washington, D.C., Dec. 1974.
[11] Hazard Analysis of Fire Extinguishers, Consumer Product Safety Commission, Washington, D.C., Jan. 1974.
[12] "The Flammable Fabrics Program 1968–1969," Technical Note 525, National Bureau of Standards, Washington, D.C., April 1970.
[13] "The Flammable Fabrics Program 1970," National Bureau of Standards, Technical Note 596, Washington, D.C., Sept. 1971.
[14] Haynes, B. W., Jr., in Proceedings: 4th Annual Meeting, Information Council on Fabric Flammability, New York, 3 Dec. 1970.
[15] Vickers, A., Krasny, J., and Tovey, H. in Proceedings: 7th Annual Meeting, Information Council on Fabric Flammability, New York, 5 Dec. 1973.
[16] Buchbinder, L. B. in Proceedings: 7th Annual Meeting, Information Council on Fabric Flammability, New York, 5 Dec. 1973.
[17] Buchbinder, L. B., Journal of Fire and Flammability—Consumer Products, Technomic Publication Company, Inc., Westport, Conn., March 1974.
[18] Annual Report, Consumer Product Safety Commission, Washington, D.C., Oct. 1974.

**Bibliography**

*Consumer Product Safety Commission*

*The Consumer Product Safety Act: Text, Analysis, Legislative History,* Bureau of National Affairs, Inc., Washington, D.C., 1973.
*NEISS News,* Bureau of Epidemiology, Consumer Product Safety Commission, Washington, D.C., monthly.

*Aluminum Wire*

Bunten, E. D., Donaldson, J. L., and McDowell, E. C., *Hazard Assessment of Aluminum Electrical Wiring in Residential Use,* NBSIR 75-677, National Bureau of Standards, Washington, D.C., Dec. 1974.
Consumer Product Safety Commission, transcript of hearings on aluminum wire, March 1974 (unpublished).

*Television Receivers*

Consumer Product Safety Commission, transcript of hearings on television receivers, April 1974 (unpublished).

*Fire Detectors*

Custer, R. L. P. and Bright, R. G., "Fire Detection: The State-of-the-Art," Technical Note 839, National Bureau of Standards, Washington, D.C., June 1974.
*Household Fire Warning Equipment: Spot Type Detectors,* Fire Equipment Manufacturers' Association, May 1974.

*Flammable Fabrics*

*Fire Safety Research,* M. J. Butler and J. A. Slater, Eds., Special Publication 411, National Bureau of Standards, Washington, D.C., Nov. 1974.

*Miscellaneous*

Marzetta, L. A., "Engineering and Construction Manual for an Instrument to Make Burn Hazard Measurements in Consumer Products," Tech. Note 816, National Bureau of Standards, Feb. 1974.
"Fire Extinguishers . . . Can They Be Dangerous?," National Association of Fire Equipment Distributors, Inc., Chicago, 1970.

# DISCUSSION

*I. N. Einhorn*[1] (*oral discussion*)—The CPSC has sponsored research by Dr. Dressler's group at Harvard University. After submission of Dr. Dressler's final report, a number of organizations were asked to critique this report. The results of this critique would be most useful to many scientists and engineers working in this area, for example to members of ASTM Committee E 5.02.02. Has there been a summary of this critique, and, if so, where can one obtain this summary?

*Walter Thomas* (*author's oral closure*)—To my knowledge, there has been no report filed on the critique by the group that were requested to prepare one. The status of this can be obtained from the Bureau of Biomedical Science, in our organization, which has the primary responsibility for this particular project.

[1] Professor, Flammability Research Center, University of Utah, Salt Lake City, Utah 84108.

*J. Witteveen[1] and L. Twilt[1]*

# International Standardization: Advantages, Progress, and Problems

**REFERENCE:** Witteveen, J. and Twilt, L., "**International Standardization: Advantages, Progress, and Problems,**" *Fire Standards and Safety, ASTM STP 614,* A. F. Robertson, Ed., American Society for Testing and Materials, 1977, pp. 253–265.

**ABSTRACT:** Four basic requirements of fire safety standards are proposed: the system of international rules must be objective, practical, mandatory, and consistent. Taking into account the state of the art, the fulfillment of these basic conditions is subjected to some substantial constraints. A very important constraint appears to emerge from the inconsistency between the quantification of levels of protection and the quantification of measures to meet the required levels. Other constraints follow from political and legal considerations. A discussion on these constraints is presented, and recommendations are given to face the identified problems.

**KEY WORDS:** fires, fire safety, standardization, regulations, standards, codes, fire test

In contrast to other phenomena, even in ancient times, the occurrence of fire was considered to be a disaster, the frequency and effect of which could be reduced by rational measures. Due to the restricted knowledge and technical possibilities, however, several fires have occurred which not only led to extreme losses but, in fact, threatened the continuation of whole communities. An example of such a disaster is the severe fire which destroyed the main parts of Rome during the reign of Emperor Nero. Nero officially blamed the Christians for it. The authorities who were responsible for the rebuilding of the city, however, knew better. So they stipulated rather detailed directives on, for example, the width of the streets, the materials to be used in facades, how to organize activities involving extreme fire hazards, etc.

More recent history also provides a number of illustrations of fires which resulted in an almost total burnout of whole cities. Especially in the middle ages, such disasters occurred rather frequently. Some cities are even reported

---

[1] Deputy director and research engineer, respectively, Institute TNO for Building Materials and Building Structures, Rijswijk, Delft, The Netherlands.

to have undergone this fate several times. Commonly, such fires led to a revision and a reinforcement of the existing directives on fire safety.

As a result, the grip of the authorities on fire safety gradually became stronger. The first written directive on fire safety for a city like Amsterdam, for example, is dated 1399. This directive was incorporated in the regular system of law, valid within the city. It was comprised not only of directives on fire safety but also a detailed system of punishment in case the given rules appeared to be neglected. And this punishment was extremely severe [1].[2]

In the beginning, due to the autonomy of small cities and counties, the directives were only valid for rather insulated areas. This insulated development led to important differences in the several approaches. The growing tendency to national unities, which became manifest in the nineteenth and in the first part of the twentieth century led to a certain harmonization of these directives on a national scale, although, in many countries, local differences still exist. This development, which is laudable in itself, consequently gave rise to a rather strict and formalized system of rules, in many cases, cramped into the national legislation. Consequently, the existing differences between the several systems were sharpened.

Conditions today, however, are characterized by rapidly increasing communication and trade. As a result, industrial and other activities are no longer restricted to national borders.

Differences in national approaches obviously frustrate such a development. Consequently, the several national systems should be harmonized and replaced by an international one.

## Main Features and Basic Conditions

The goal of standardization on fire safety is to create an objective base on which both (*a*) the levels of protection against fire and (*b*) the technical measures to meet the required levels can be quantified.

International standardization means that the several national systems concerned with fire safety in buildings, which, for many reasons, conflict, must be harmonized, taking into account, of course, the just-mentioned goals. The specific aim of such standardization is to level the existing differences where they are irrelevant.

Before discussing, in more detail, the ways and means to arrive at an international system, it is necessary to describe more precisely what it implies and which basic conditions are to be taken into account. In this respect, the following terminology will be used.[3]

*Regulations*—These are documents, having force of law, detailing pur-

---

[2] The italic numbers in brackets refer to the list of references appended to this paper.

[3] The terminology presented here does not necessarily coincide with the terminology used in the respective countries. In this respect, also, standardization of terminology will be useful.

poses and listing general requirements on levels of protection; such regulations commonly function as a framework for more specific requirements.

*Codes*—Codes signify documents, referred to in the regulations and comprised of more specific requirements on levels of protection; such codes are valid for very specific types of buildings (for example, hospitals and schools). For practical reasons, sometimes practical solutions to meet required levels of safety are also included in codes; these are called codes of practice.

*Standards*—These documents are referred to in the regulations and codes and are comprised of specific techniques and methods which are necessary in order to quantify the measures to meet the required levels of protection.

*Standardization*—Standardization is the activity necessary to arrive at a consistent set of regulations, codes, and standards with respect to fire safety in buildings.

In the subsequent discussion, regulations, codes, and standards will be designated as rules.

In Fig. 1, an outline is given concerning the structure of regulations,

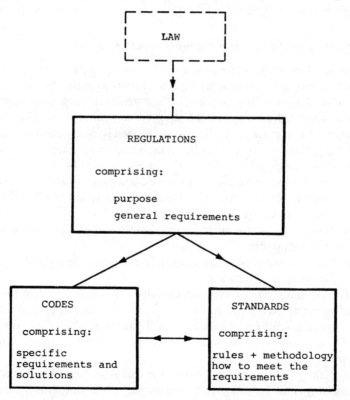

FIG. 1—*Schematic presentation of the structure of regulations, codes, and standards on fire safety of buildings.*

codes, and standards (rules) which are common practice in many countries. Harmonizing the several national systems into an effective international system obviously necessitates the harmonization of all mentioned components. In doing so, the following basic conditions are to be taken into account.

1. The system must be objective. This yields to the suggestion that the system is to be based on functional requirements, rather than on typological description of products.

2. The system must be suitable for practical use. This yields to the necessity of simplification of the rather complex reality behind the system.

3. The system must be mandatory. To ensure general application of the system, it must have the force of international law.

4. The system must be consistent. As has been stated before, the intended goal of standardization is twofold: quantification of both the required levels of protection and the measures to meet these required levels. Obviously, there must be some balance in the way in which these two items are prescribed, since it is pointless to reach for a high level of perfection in one field if the other field is hardly elaborated.

### Suggested Ways to Achieve International Standardization

International rules must be based on functional aspects. In analyzing these aspects, two main items are of interest: (a) the probability and nature of fires in buildings and (b) the material and structural response to the fire. Therefore, in setting up a system of international rules, it is logical to use these aspects as a starting point. Such an approach is indicated in Fig. 2.

First of all, these aspects are to be established in as complete a way as possible. This is essentially a statistical and physical problem.

The next step must be an evaluation of these aspects to arrive at: (a) performance criteria (formulated in a broad sense) and (b) methodology to predict the material and structural behavior during fire.

The performance criteria at this stage should be formulated in a rather broad sense. For example:

1. "The structure has to withstand the fire during a period long enough to permit the occupants to escape."

2. "The structure has to withstand a fully developed fire, and the condition of the structure must be such that, after extinguishing, economic repair is possible." Which of such criteria are appropriate in a certain situation or for a certain type of building must be determined on the basis of socioeconomic considerations.

Based on insights concerning the material and structural response to fire, a methodology must be designed by which the actual behavior during fire can be predicted. Such methodology can be based on tests as well as on calculations, depending on the state of knowledge.

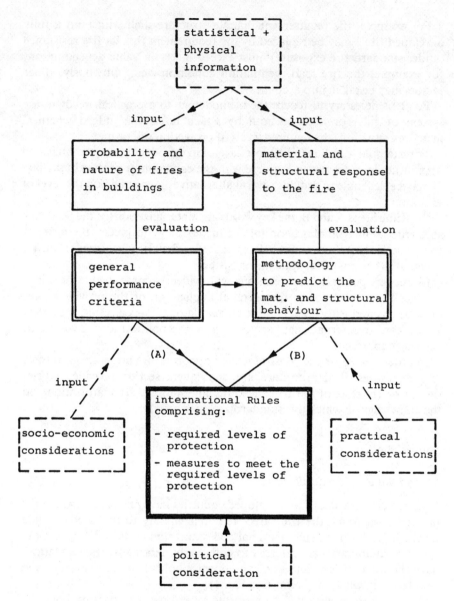

FIG. 2—*Standardization on fire safety in relation to functional aspects.*

At this stage the following steps must still be made.

A.    Transformation of the performance criteria to levels of protection must be accomplished. Performance criteria formulated in a broad sense are to be quantified by transforming them into well-defined criteria (levels of protection).

For example, the requirement that "a structure shall withstand a fully developed fire" may be replaced by the requirement that its fire resistance (under standard fire exposure) must exceed a certain value depending on, for example, the fire load, ventilation conditions, etc. Obviously, other factors may be taken into account.

B.   It is necessary to reduce the methodology to a practical, ready-to-use system of rules or data or both by which it can be judged whether, in a given situation, the required levels of protection will be met.

For example, one has to be most careful in introducing very complicated approaches in order not to interfere with the condition of practical usability. It may be desirable to provide several alternatives, depending on the level of ambition.

Realizing Steps A and B, the key words are materialization (of the performance criteria) and simplification (of the methodology to predict the material and structural response to fire). Both activities shall be performed in such a way that the extent of reliability is consistent.

Moreover, political considerations will evidently come to the fore since another basic condition is that general application of the provided rules must be ensured ("the system must be mandatory"). As a result, especially with regard to international harmonization, the political aspect will be of extreme importance.

In order to analyze the constraints and opportunities present with respect to the just-described procedure, the subsequent section provides a brief outline on the state of the art concerning international standardization and the knowledge on which this standardization is based.

## State of the Art

### Present State of Standardization

At present, practically all developed countries have regulations where the principles on which the fire safety requirements are to be based are laid down. In such documents, national codes and standards which give more specific requirements are referred to commonly. Normally, the regulations have the force of law, and this consequently holds true for the codes and standards to which they refer (see Fig. 1).

It appears, however, that, between the several national systems, considerable differences do exist. As a result, international standardization is far from being at hand. Looking at the actual sources from which the differences emerge, the following distinctions can be made: (a) philosophy, (b) technical aspects, and (c) legal aspects. Subsequently, a brief review of the three mentioned items will be given.

As far as philosophy is concerned, in practically all countries, requirements on fire safety in buildings are based primarily on safeguarding against loss of life and injury of the occupants. It is thus generally required that structural integrity is maintained whereby occupants can escape. In addition, in many countries, it is required that measures be taken so that the fire will be limited to one building or to one compartment. The latter point gives rise to some uncertainty since, in many cases, it is not clear whether reduction of economical losses should also be recognized as a main goal of the regulation on fire safety. In some countries, (for example, the Netherlands [2]), the economical aspects are explicitly excluded. On the contrary, in other countries (for example, Switzerland), the economic aspect is fully incorporated in the rules [3]. In the latter case, a very important role is played by the insurance companies. Considering the several national approaches, one very striking point is that the relationship between the goal of the regulations on fire safety and the actual levels of protection is seldom, if ever, specified. It is thought that a crucial obstacle on the way to international standardization emerges from this discrepancy.

As far as the technical aspects are concerned, the situation is much better on first sight. In this field, international organizations like International Standards Organization (ISO) have already been active for many years. The work done by ISO/TC 92 must be mentioned in this respect. A good illustration of this international cooperation is the "Recommendation on Fire Resistance Tests of Elements of Building Constuctions" (ISO/R 834) [4]. Most of the countries do accept the major requirements of this document. It must be kept in mind, however, that, on many details, a considerable divergence between ISO/R 834 and national standards still exists [5,6].

For example:

1. The failure criteria for fire-exposed structural members are defined in a rather general way. As a result, in the national standards, additional rules are given, which unfortunately appear to differ considerably from each other.

2. In most of the countries, the ISO fire curve is adopted as a conventional condition as far as the ambient gas temperature is concerned. However, the other conditions during the fire tests (for example, measurement of temperatures, dimensions and number of specimens, conditioning, load and restraint conditions) still display discrepancies.

In this respect, it must be mentioned that most of the countries only accept a classification of materials or elements based on fire tests. However, in some countries, the possibility for a more theoretical approach is available. An illustration of this is given in the Swedish regulations [7]. Based on these regulations, it is possible in Sweden (as an alternative to more conventional methods) to analyze the behavior of fire-exposed steel structures under natural fire conditions by means of calculation [8]. In this context, however, attention must also be paid to the initiative of international material organizations like Comité International du Béton (CEB), Fédération International de la Précontrainte (FIP), and Convention Européenne de la Construction

Métallique (CECM). The first two organizations recently issued a first draft for international recommendations with respect to the design of reinforced and prestressed concrete sturctural members for fire resistance [9]. Within the CECM, activities are in progress to arrive at analogue recommendations in the field of steel structures.

Summarizing, it must be stated that, although considerable progress has been made in harmonizing the technical aspects, the present situation is not yet quite satisfactory. As a result, in Europe, for example, considerable difficulties concerning recognition and exchange of test results of the different national laboratories are encountered. An effort on the part of the Commission of European Community to alter this situation must be mentioned. Also, European stations on fire research have formed a group which deals with this problem (International Laboratory Data Acceptance (ILDA)).

Concerning the legal aspects, it can be noted that, in most countries, standardization on fire safety is based on national legislation. It must be kept in mind, however, that, in some cases, harmonization is far from being achieved even within the national borders. In Great Britain, for example, separate regulations on fire safety are in force for England and Wales, Scotland, Northern Ireland, and Inner London Area. All these regulations differ on the required levels of protection [10]. Such situations, partly resulting from local circumstances, partly stemming from the past, are present in many other countries. They obviously interfere with efforts toward achieving international standardization. Another complication results from the fact that the more specific stipulations (codes, see Fig. 1) are generally only given for a limited number of building types (for example, high-rise flats, hospitals, schools, etc.). The building types for which rules are given do not coincide in the different countries. If the building is not covered by the available codes, establishing the protection level is generally left to the competence of the local authorities. In a country like the Netherlands, this responsibility is given to the municipal government. In other countries, it may belong as well as to the government of the state or the province. As can be expected, this situation sometimes leads to important and unjustified local differences between required levels of protection even within one country.

## Present State of Knowledge

To meet the intended goals of standardization on fire safety (see the section on basic standardization conditions), a considerable amount of knowledge has to be assembled. In the last few decades, this knowledge has increased substantially as far as the understanding of the fire process itself and the structural response to fires is concerned [11-12]. In this respect, once again, the valuable efforts of international material organizations like CEB, FIP, and CECM must be mentioned (see, for example, Ref. *13*). Also the Comité International de Bâtiment (CIB) is active in this field, with special emphasis on international exchange of knowledge.

Obviously, this knowledge can be used to derive and refine methods (both theoretical and experimental) on ways of meeting given levels of protection. However, the problem of how to establish realistic levels of protection cannot be solved simply by applying such knowledge. Economic and social aspects must be taken into consideration to achieve this need. Apart from some pilot studies [*14-16*], however, no attempts have been made so far to extend our knowledge in this field. It is thought that this imbalance is mainly caused by a general preference of the scientist for physical quantification. Phenomena which cannot be expressed in physical or statistical terms and thus cannot be quantified by conventional means are, to some extent, out of the scientist's field. Von Goethe, in his famous play, *Faust*, makes mention of this feature, putting the following words in the mouth of Mephistopheles[4]: [*17*]

> There spoke the veriest bigot of booklearning,
> What you discern not, Sir, there's no discerning
>
> All, that you touch not, stands at hopeless distance
> All, that you grasp not, can have no existence.

As will be elucidated in more detail in the next section, it is felt that the identified gap in knowledge represents one of the main points of weakness in the present situation with respect to any attempt to reach for an international system of rules on fire safety.

## Constraints and Recommendations

The fulfillment of the four basic conditions formulated in the beginning of this paper is subjected to some substantial constraints. In this section attention will be paid to these constraints. Moreover, recommendations will be given to face the identified problems.

It is thought that important problems will be encountered in trying to fullfill the condition of consistency between: (*a*) quantification of levels of protection and (*b*) quantification of measures to meet the required levels.

As pointed up in the section on the state of knowledge practically nothing has been done so far concerning the first item. Emphasis has been on the second item, however. It is felt that this situation, in the end, will have very serious and unfavorable effects: the existing inconsistency between the choice of levels of protection and the quantification of the

---

[4] Adapted from the German [*18*]:

"Daran erkenn ich den gelehrten Herrn!
 Was ihr nicht tasted, steht euch meilenfern,
 Was ihr nicht fasst, dass fehlt euch ganz und gar
 Was ihr nicht rechnet, glaubt ihr, sei nicht war.''

measures to meet these levels will widen. As a result, the efforts in the technical field will degenerate more and more into attempts to convince the responsible authorities, adhering to the letter of the law. Due to its exactness, a technical solution may suggest a great reliability on the measures on which it is based. However, as long as this inconsistency has not been eliminated, such suggestions are misleading and may even be dangerous.

Consequently, the quality of the standardization will be endangered. Moreover, basic differences in the national approaches do exist with respect to the required levels of protection. However, a sound base on which an international discussion on this matter can be carried out is lacking. It is felt that such a base is a necessary condition for international agreement. It is strongly recommended that primary efforts in the study be directed toward attaining desirable and motivated levels of protection. In this discussion, both scientists and authorities responsible for fire safety should participate. An effort on the part of CIB, W14, to deal with this problem must be mentioned (Code Advisory Panel). This implies an extension of the terms of reference of CIB, W14.

A further constraint follows from an essential discrepancy between some of the basic conditions. The condition of reaching for an objective system, on one hand, urges us to a functional approach. As a result of the complexity of the physical, social, and economical mechanisms involved, such an approach will lead to rather detailed and subtle rules. On the other hand, however, conditions concerning the practical usability and the mandatory character of an international system of rules urge us in the direction of simplification. Political considerations will also force a trend to rather simple and not very differentiated solutions, since only common ideas will be attainable. For example, in the last few decades, substantial progress has been made with respect to the state of knowledge about natural fires and structural response to fire. However, this increasing state of knowledge, commonly recognized, has been reflected only partially in the national standards. Therefore, it is doubtful whether, from a political point of view, an international agreement on such unconventional approaches will be possible. For the several reasons just mentioned, it is felt that important simplifications must be accepted. It is pointed up, however, that an explicit motivation and reference to the background of the simplified rules are essential.

An often-mentioned type of constraint is introduced by the occurrence of local differences. Of course, one has to realize that such differences do exist (for example, with respect to equipment of the fire brigade, methods of construction, density of the population, etc.). However, in the opinion of the authors, local differences are sometimes used incorrectly as arguments to retard harmonization.

Closely related to the political aspect is the legal aspect. Since a basic condition is that each system of rules needs to have force of law, such a system is likely to have repercussions with respect to the national legislation.

For this reason, it seems only sensible to reach for international standardization if the need for harmonization is supported by the intergovernmental willingness to abandon national conceptions. In this respect, it is very strongly recommended to involve legal considerations right from the beginning.

## Conclusions

The authors are aware of the fact that this paper is not a blueprint for international standardization. More problems have been raised than solved. However, it is felt to be extremely important that a discussion on the problem, taking into account all aspects, has begun. The aim of this paper is basically to stimulate this activity.

## References

[1] Verburg, G. J., "Brandpreventie vroeger" ("Fire Prevention in Former Days"), *De Brandweer,* Vol. 28, No. 8, Aug. 1974.

[2] Richtÿnen Brandbeveiliging van Gebouwen -Hoge woongebouwen-'' ("Directives for Fire Protection of Buildings—High-Rise Flats"), NEN 3893, Nederlands Normalisatie-Instituut, Rijswijk (Z.H.), Nov. 1975.

[3] Wijss, U. in "Fire Safety in Buildings with Steel," 1st International Symposium, Convention Européene de la Construction Métallique Symposium, 74-1 FED, Den Haag, Oct. 1974, pp. 48–56.

[4] "Fire Resistance Tests of Elements of Building Construction," ISO 834, International Standards Organization, 1974.

[5] Dekker, J., "Differences between ISO/DIS 834 and National Standards," Institute TNO for Building Materials and Building Structures Report B VI-74-65, Oct. 1974.

[6] Malhotra, H. L., "Preliminary Report on the Response to the Questionnaire on Loading Restraint and Deflection Criterion," ISO/TC 92/WG 5 (UK 5), No. 34, International Standards Organization, 1974.

[7] Svensk Byggnorm 1967 (SBM 67), (Swedish Building Regulations 1967 (SBR 67)), Statens Planverk, Publikation No. 1, 1967.

[8] Magnusson, S. E., Petterson, O., and Thor, J., "Brandteknisk dimensionering av stalkonstruktioner," ("Fire Engineering Design of Steel Structures"), *Stålbyggnadinstituted, Publikation 38, 1974.*

[9] "Guides to Good Practice," FIP/CEB Recommendations for the Design of Reinforced and Prestressed Concrete Structural Members for Fire Resistance, FIP/1/1, 1st ed., Federation International de la Précontrainte/Comité International du Béton, June 1975.

[10] Eliot, D., *Building with Steel*, No. 19, Feb. 1975, pp. 4–10.

[11] Thomas, P. H., "Fires in Enclosures," CP 30/74, Building Research Establishment Current Paper, Feb. 1974.

[12] Bobrowsky, J. et al, *Fire Resistance of Concrete Structures,* Institution of Structural Engineers, London, Aug. 1975.

[13] "Fire Safety in Structural Steelwork," European Convention for Constructional Steelwork, CECM-III-74-2E, Convention Européene de la Construction Métallique, 1974.

[14] Witteveen, J. and Twilt, L., in *Fire Safety in Constructional Steelwork,* European Convention for Constructional Steelwork, CECM-III-74-2E, Convention Européene de la Construction Métallique, 1974, pp. I.1--I. 66.

[15] Melinek, S. J., "A Method of Evaluating Human Life for Economical Purposes, "Fire Research Note No. 950, Fire Research Station, United Kingdom, Nov. 1972.

[16] Lie, T. T. in *Fire and Buildings,* Applied Science Publishers Ltd., Barking, England, 1972, Chapt. 9, pp. 211–239.

[17] Anster, J., *Second Part of Goethe's Faust*, London, 1835.
[18] Goethe, J. W. von, *Faust, Der Tragödie erster und zweither Teil*, Atlas-Verlag-Koln, p. 125.

# DISCUSSION

*R. M. Fristrom*[1] *(oral discussion)*—International standardization of fire tests and codes is very important. We have an even more complex problem in the United States because of many local jurisdictions. Could you comment on the relationship between your international efforts and our own efforts to standardize within this country? You are interested in European aspects: do they have bearance on American standards?

*J. Witteveen (author's oral closure)*—This is a very interesting question because, if you want to have international standardization, of course, you have to begin with your own country. If you look to even a small country like the Netherlands with about 15 million inhabitants, standardization is not at hand at the moment. There are many local differences, especially as far as the levels of protection are concerned. As I learned from my American colleagues, in the United States you have the same problems many other countries have. But I think it is esstenial that we try to have an international discussion on a scientific and logical basis. This should also be used for stimulating similar activities in our own countries.

*H. Tovey*[2] *(oral discussion)*—One of the problems that we run into, for example, in data processing where the uniformity of procedures is terribly important, is the incentive we can offer the prospective participants to persuade them to change their way of doing things. There is really not much that we can offer. We can say: we want to compare our data with yours. But they may not care. I am afraid that we will have to live for a long time before uniformity is greatly improved. I would like to suggest something that you did not address in your talk, namely, that we ought to start thinking about how to live with different systems. What methodologies can be developed to permit us to make some correlations using different data or data collected on different bases.

*J. Witteveen (author's oral closure)*—Of course, international standardization is a fairly long-term exercise, and, for the time being, we shall have to live with the different systems. Your suggestion about correlating the different systems can be used and is, in fact, followed today, as far as the technical measures to meet required levels are concerned. In my opinion, this will, in say ten years, converge into an international standardized

[1] Senior staff chemist, Applied Physics Laboratory, The Johns Hopkins University, Silver Spring, Md. 20910.
[2] National Fire Data Center, National Fire Prevention and Control Administration, U. S. Department of Commerce, Washington, D. C. 20234.

system (for example, tests, calculations).  As far as the quantification of levels of protection is concerned, there is a need for an international accepted methodology on which the national approaches can be based. This activity will take a much longer time than the ten years I mentioned before.  Until now, the national approaches have not been consistent and are sometimes illogical, so that the systems cannot be correlated.

*R. M. Fristrom*[1]

# Sampling and Analysis of Fire Atmospheres*

**REFERENCE:** Fristrom, R. M., **"Sampling and Analysis of Fire Atmospheres,"**
*Fire Standards and Safety, ASTM STP 614,* A. F. Robertson, Ed., American Society
for Testing and Materials, 1977, pp. 266–284.

**ABSTRACT:** The characterization of fire atmospheres is discussed. A complete
description would require more than $10^{17}$ measurements. This is beyond present
capabilities, but the requirements could be reduced by separating the problem into
three regimes, each characterized by a minimum volume-time ($v$, $t$) increment. They
are: (*a*) overall fire atmospheres ($v = 1$ m$^3$, $t = 10$ s) for hazard and detection,
(*b*) fire plumes ($v = 10^{-6}$ m$^3$, $t = 10^{-1}$ s) for heat transfer and fire propagation, and
(*c*) flames ($v = 10^{-15}$ m$^3$, $t = 10^{-4}$ s) for microscopic and molecular combustion
processes. By systematically applying these characterizations to the regions where they
are required, the data base can be reduced to $10^6$ measurements for a typical small
fire. This is feasible with present technology. However, the application would require
a substantial investment in facilities and the development of sampling and analytical
technique. Some sampling and analytical problems associated with characterizing
fire atmospheres are discussed. The current status is illustrated using several recent
studies.

**KEY WORDS:** fires, analysis, flames, sampling, measurements, fire detection systems, fire hazards, heat transfer, flame propagation, combustion

The complete characterization of a full-scale fire is a massive problem
which is impractical even with present day technology. As an example,
consider a small house fire with a volume of 1000 m$^3$ burning for an
hour. Some flame phenomena occur in random regions as small as
$10^{-15}$ m$^3$ and require a time resolution of $10^{-4}$ s. Full characterization
requires temperature, gas velocity, and the concentration of as many as
100 species in each of these minimal volume-time elements. If we assume
complete characterization of each minimal element, more than $10^{17}$ mea-

[1] Chemist, Principal Staff, Applied Physics Laboratory, Johns Hopkins University, Silver
Spring, Md. 20910.
*Supported by National Fire Prevention and Control Administration through the National
Science Foundation Research Applied to National Needs Program under Grant No. GI-
44088X.

surements would be required. A fast machine can read $10^3$ measurements/s and would take three million years to read the data. No machine approaches this storage capacity. The problem is also beyond present analytical technology. To obtain the data in real time would require ten thousand million separate measuring devices. This would disturb the system strongly. If every chemist in the U.S. did 3000 analyses/s/day it would require 3000 years to complete the project. Few analyses can be completed in a second. The Orsatt analysis and titrations are not candidates for this job. Even assuming that the data could be obtained, it is clear how such a flood of information could be analyzed.

One systematic approach to the problem is to isolate the critical information and divide it into manageable segments. Much of the total information available is redundant or unnecessary. During the early history of the fire, only the region around the ignition requires characterization. High time spatial resolution is required for combustion waves, but they persist only for short periods, and much of this behavior is better studied in the laboratory. Considering the hierarchy of scales and using those appropriate to the processes under study, one might divide a fire into volumes 1 m on a side characterized at 10-s intervals over a 1-h period. This would require $10^6$ measurements, and a single multiplexed high-speed data channel would be adequate with video tape storage. The next lower structural heirarchy would be volumes 1 cm$^3$ with time resolution of $10^{-1}$ s. Only a few of the original 1 m elements would require this detailed characterization, but the locations requiring this detail change as the fire progresses. The information required is temperature and flow patterns in and around the progressing fire. Specification at this level might require $10^8$ measurements. This is equivalent to ten video channels. Data handling at this rate has been done in aerodynamic testing [1][2] but never for hour-long periods. This amount of information would be difficult to interpret.

The finest scale processes are those occurring in the flames. The spatial and time resolution requirements make the use of sophisticated instrumentation mandatory. This is much better done under laboratory rather than field conditions. Since these microscopic combustion processes depend on local conditions rather than overall fire characteristics, this is a feasible separation. The number of potential fuels is large, and the oxygen level can vary from normal air at 20 percent to almost any level of vitiation. In addition, the radiation level and turbulence level are important parameters. To cover all of these local conditions would require a study of perhaps 100 fuel types at five oxygen levels, five flow conditions, and five radiation levels. This would require a bank of at least $10^9$ measurements, but it would have the virtue of being generally applicable to fire

[2] The italic numbers in brackets refer to the list of references appended to this paper.

situations and reusable. Only a small fraction of this information is available in the present literature.

Even this program is too ambitious for our present resources. The facility which could execute this problem over a period of a decade would require an investment of 10 million dollars a year over ten years. This is not beyond society's capabilities but is unlikely in the present economic-political climate. Therefore, we turn our attention to what has been done and what might be possible.

A well-instrumented room burn will utilize perhaps 100 thermocouples operating at a 10-s repetition rate, 50 velocity sensors operating at a 10-s repetition rate, and a dozen gas sampling points operating at a 100-s repetition rate. A typical run time might be 1000 s. This results in the accumulation of $2 \times 10^4$ measurements which is a respectable amount of information to interpret. One can visualize improvements by two orders of magnitude as microminiaturization, microprocessors, and microcomputers are applied to instrumentation. This should make instruments cheaper with increased channel repetition rates and allow direct transcription of data to the computer. This approaches our idealized characterization so that one might hope that, with normal improvements, good statistical design, and intelligent experiments, a meaningful "complete" description of a fire could become a reality in the next decade.

**Sampling Problems**

The description of a region requires specification of temperature, pressure, gas velocity, and composition. Since composition is a manifold variable while temperature is a single scalar, and velocity a three-component vector, the bulk of information is composition. We will therefore discuss principally composition specification, which is the most difficult problem and in need of improvement. It should be remembered, however, that temperature and velocity are basic to the problem. Pressure is usually constant.

*Sample Requirements*

A meaningful sample must represent a known average of the particular time-volume element being withdrawn. Composition, temperature, and gas velocity vary with position and time as the fire evolves. Large-scale variations occur because of stratification, flow, and fire progress. Fine-scale variations result from turbulence and reaction processes. On the microscopic level, this problem is best solved by following the changes in real time. On the intermediate scale, one is interested principally in flows, turbulence levels, and temperature. The principal composition information required is the fuel-oxidizer level which can be used to determine local combustion potential. On the macroscopic scale, the averaging problem is severe because most probes characterize a region which is small compared with the volume to be characterized, and variations over the volume are often severe and significant. These problems are outlined in more detail in the sections on probes and sample averaging.

A second important question is the level at which an analysis should be terminated. In a problem as complex as fire atmosphere characterization, it is unrealistic to ask for a complete analysis. As analytical sensitivity is increased, the number of detectable species goes up. With parts per million as a minimum detectable level, hundreds of compounds can be detected in a typical fire. Present day attainable levels which go as high as one part in ten billion would probably increase this a hundredfold. What is needed is a rational criterion for setting the minimum level of detection. For the fire problem, this criterion is the concentration at which a dangerous short-term effect on people or property exists. Since people are usually more delicate than property, limits should be related to life safety. Fire exposures are usually fractions of an hour. Therefore, the ideal is to determine the concentration and distribution of those species which adversely affect people in these concentration-time dosages. For materials such as the phosphoryl compound found by the Utah group [2], this is $\sim 10^{-6}$, [3] but for most toxicants, the dangerous level lies above $10^{-3}$ (Table 1). Soot particles, liquid droplets, and inorganic dust are also produced in fires. Polar gases are adsorbed on particle surfaces. This presents a complex analytical problem.

TABLE 1—*Toxic products of combustibles.* *

| Toxic Product | Common Source | Free Burning | Smoldering |
|---|---|---|---|
| Oxygen depletion | combustion | 0 | + |
| Carbon dioxide | carbonaceous materials | + | − |
| Carbon monoxide | carbonaceous materials | − | + |
| Nitrogen oxides | celluloid, polyurethanes | − | + |
| Halogen acids | fire-retarded plastics, halogenated plastics | + | + |
| Ammonia | melamine, nylon, urea-formaldehyde resins | 0 | + |
| Sulphur dioxide | rubber, thiokols | + | − |
| Hydrogen cyanide | wool, silk, plastics containing nitrogen | − | + |
| Benzene | polystyrene | 0 | + |
| Phenol | phenol-formaldehyde resins | − | + |
| Aldehydes | wood, paper | − | − |
| Organic acids | cellulosic materials, rayon, nylon, polyesters, phenol-formaldehyde resins | − | + |
| Phosgene | chlorinated plastics | 0 | − |

NOTE—
+ Dangerous concentration.
− Small concentration.
0 Negligible concentration.
* Modified after Wagner [4].

*Chemical Species in Fires*

A fire atmosphere is a complex chemical mixture with gas phase species and liquid and solid particles.

Gas phase molecules can be classified as stable species, with life times which are long compared with sampling and transfer times, and short-lived unstable species. In most parts of a fire, the species are stable. The flame region, however, involves high temperatures where short-lived molecular fragments such as radicals, atoms, and ions are produced. They are best studied in the laboratory where special instrumentation is used for rapid sampling. Flame studies have been reviewed recently [1,5]. Fire gases can contain a host of toxic species; the most important are listed in Table 1.

*Probes*

Fire atmospheres are usually analyzed by withdrawing a sample through a probe and transferring it to an analytical system. Some typical probes are shown in Fig. 1. Two types of probes are used, the sonic microprobe, Fig. 1c, and the cooled subsonic probe, Figs. 1a,b. The design of probes was reviewed by Tinè in 1961 [9] and by Bilger [6] in 1973. A recent discussion of microprobe sampling in flames has been given by Fristrom [1,5].

Sonic probes sample gas phase composition directly and allow determination of radical species which are important in flames. Problems occur with this probe type if particulates are present, as occurs in many fires. Microprobes are easily clogged and small particles are sampled preferentially. A subsonic probe sampling at local stream velocity (called isokinetic sampling) allows an unbiased sampling of particle ladened atmospheres. However, quenching of the sample is accomplished by conduction to the probe walls, and this is often inadequate for rapidly reacting flame gases. In addition, sampling flows of varying velocity can induce acoustic oscillations which disturb the sampling rate.

Spatial and time inhomogeneities lead to errors both when they are small compared with the volume to be characterized and when they are small compared with the probe sampling volume. The ideal situation would be to sample a small enough region with a short enough sampling time so that the sample will be homogeneous in space and time. As we have seen, this leads to an impractical number of samples. This problem is not always addressed in fire studies, but it is serious. Three approaches suggest themselves:

1. Accept the averaged samples and try to estimate the error level; this is often impractical.

2. Use an analytical method which averages over a large volume [6]. This is of limited applicability because fire atmospheres are often too

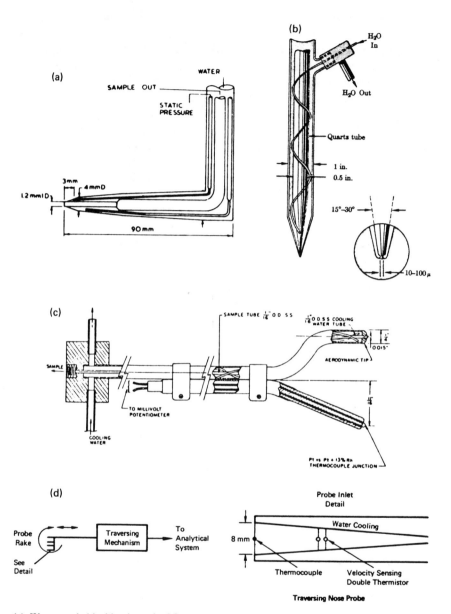

(*a*)  Water-cooled isokinetic probe [*6*].
(*b*)  Water-cooled aerodynamic nozzle probe [*7*].
(*c*)  Microprobes for flame studies [*8*].
(*d*)  Proposed traversing nose with probe and inlet flow adjusted to simulate breathing. Probe rake is traversed so that a cylindrical volume is sampled uniformly.

FIG. 1—*Sampling probes for fire atmospheres.*

smoky for transmission, and even clear atmospheres require knowledge of the temperature distribution.

3. Move the sampling point systematically through the test region during the sampling period. This approach would appear to be promising since it allows control of the averaging.

Bilger gives an excellent discussion of sample collection aerodynamics [6]. Sampling nonhomotropic gas flows (flows of nonuniform density) is discussed by Tinè [9]. Density variations in combustion can exceed tenfold over short distances. This introduces no error in isokinetic sampling (that is, the velocity into the probe is the same as the stream velocity). If the sampling velocity differs from stream velocity, however, the sample will be biased in favor of the more dense portions of the sample. The situation is further complicated by the dependence of the sampling on the acoustic impedance of the sampling line. This can be serious for subsonic sampling [9]. Near isokinetic conditions moderate velocity variations in sampling are not serious, but the shape of the probe head can have serious effect [6].

In fires, the sample will usually vary in composition during the sampling periods. This is a problem because even if the sample is not biased by the probe, it will not represent a linear time average. The averaging problem has been addressed by Bilger [6] who concludes that, with isokinetic sample collection, the sample composition should be the Favre mean. The difference between the Favre mean and the time mean can be up to 40 percent

$$Y_{i,s} = \overline{\varrho U_s Y_i}/\overline{\varrho U_s} = \overline{\varrho Y_i}/\overline{\varrho} \equiv \tilde{Y}_i \qquad (1)$$

where

$Y_i$ = mass fraction of species $i$,
$Y_{i,s}$ = sample value as analyzed,
$\varrho$ = density,
$U_s$ = sampling velocity, and
tilde = a Favre average defined as shown or in terms of the Favre probability density function [6].

With sonic orifice sampling without aerodynamic bias, the sample composition will be

$$Y_{i,s} = \overline{C_D \varrho^* c^* Y_i}/\overline{C_D \varrho^* c^*} \qquad (2)$$

This lies between the true mean and the Favre mean. Here $c^*$ and $\varrho^*$ are the sound speed and density at the throat and $C_D$ the discharge coefficient. Throat Reynolds numbers are around 100 so that $C_D$ can vary.

*Sample Transfer*

Fire atmosphere samples pass from the probe through a sampling line to a container or analytical instrument. Gas samples can be biased by adsorption or condensation of species such as water, halogen acids, halogens, hydrocyanic acid (HCN), high boiling organic molecules, ammonia, etc. Particulate material can be biased because particles of various sizes are separated by velocity gradients, or particle size may change by agglomeration. Particles may be lost on the walls of the transfer tube and species may be exchanged with the gas phase.

If one is trying to follow composition in real time, sharp changes are masked by turbulent and diffusive mixing and preferential adsorption in the sampling lines. Corrections can be made using Laplace transform theory providing the line characteristics are measured [10].

**Analysis of Fire Atmospheres**

Even if one assumes a perfect representative sample of a fire atmosphere, the problem of analysis still remains. This has two aspects: (*a*) the analysis of the gas and (*b*) the analysis of the condensed phase. Condensed material can either be liquid with dissolved gases or solid with adsorbed gases. Gas analysis is complicated by condensible species since the temperature of the fire gases usually exceeds that of sampling lines. The condensed phase can be collected on a filter and the gas taken in a sampling bottle. Both condensed and gas phases are complex mixtures containing hundreds of measurable components.

The first question is: What are appropriate compounds and limits of detectability for the fire problem? The objective could be either to determine fire mechanisms or to establish the danger of the atmosphere to life and property. As mentioned previously, fire mechanisms involve transient reactive species which are difficult to sample, and such studies are better done in the laboratory. Therefore, we will address the problem of analyzing fire samples for danger to life and property. This simplifies the problem, since compounds which are present in concentrations below the threat level can be ignored. Unfortunately, this approach leaves one dangerous loophole, the unexpected compound which is lethal in low concentrations. There is no completely satisfactory answer for this problem. The approach of Birkey at the National Bureau of Standards (NBS) [11] is one answer. This is to screen atmospheres for unusual animal toxicity which, after detection, can be identified. This was done with phosphorous toxicants at Utah [2]. This technique falls outside the scope of the present paper.

*Gas Analysis*

Many analytical methods can be used for flame gases, and the interested reader is referred to standard works on gas analytical techniques. Major species, such as oxygen, carbon dioxide ($CO_2$), carbon monoxide (CO), and water ($H_2O$) can be continuously monitored using gas chromatographs, mass spectrometers, or optical spectrometers. These are all moderately expensive instruments, and large numbers of analyses are required for the fire characterization. There is hope of improvement of the situation through automation of analyses [12] (Fig. 2) with data reduction

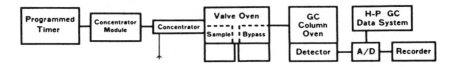

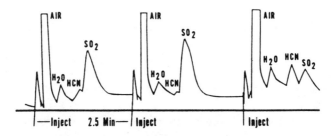

FIG. 2—*Automated gas chromatograph for analyzing pollutants* [12].

using the new generation of microcomputers [13]. These techniques have not yet been combined for the fire field. A number of complex automated clinical analyses have been developed [12] which offer a model for the fire problem.

*Condensed Phase Analysis*

Solid samples can be characterized by particle size distribution and shape. Instruments exist for making such measurements both *in situ* using scattered light and on collected samples using automated counting and pattern analysis [14] (Fig. 3). Scanning electron microscopy provides a tool which allows size and shape determination and elemental analysis.

The identification of the material adsorbed on the soot particles is more

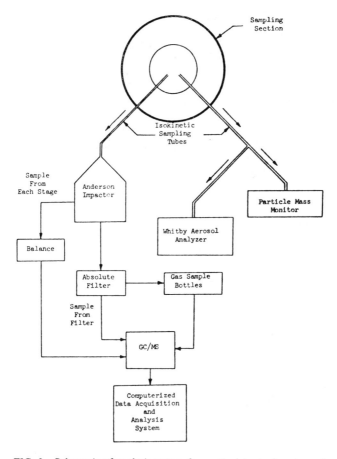

FIG. 3—*Schematic of analysis system for particulates in fire atmospheres.*

complex and is simplified by separation into classes followed by liquid chromatography [*15*] (Fig. 4).

## Examples

Studies of fire atmospheres are usually less ambitious than the total approach previously discussed. A typical study has a few well defined objectives which require the determination of only a few components. Common choices have been oxygen, carbon dioxide, carbon monoxide, with one or two toxic species (under special circumstances) such as hydrochloric acid (HCl) in the burning of chlorinated polymers, or HCN in the burning of wool or polyurethane, etc. We will give several examples of recent studies with short comments and close the paper with the discus-

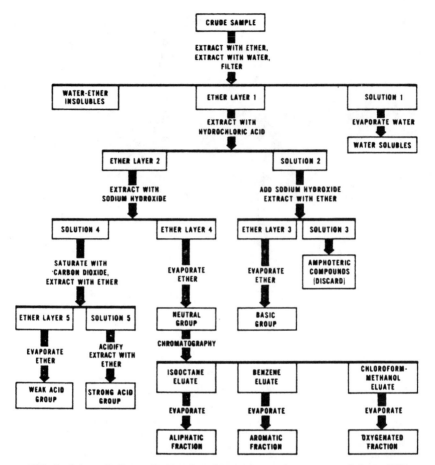

FIG. 4—*Schematic for separation of smoke constituents into compound classes* [15].

sión of an idealized system which might be constructed using present technology.

### Fire Atmosphere in an Upstairs Bedroom During Downstairs Fire

The intent of this study was to establish the buildup of life hazard in an upstairs bedroom during a downstairs fire [16] (Fig. 5). Two fire sources were used: (*a*) a fast burning wooden crib and (*b*) a slow burning sofa. Oxygen, carbon dioxide, and carbon monoxide were measured at three heights. Samples were taken by a diaphragm pump, stored in glass sample bottles, and analyzed in the laboratory by gas chromatography.

### Fire Atmosphere in a One-Story Building

Measurements were made of the carbon dioxide and carbon monoxide build-

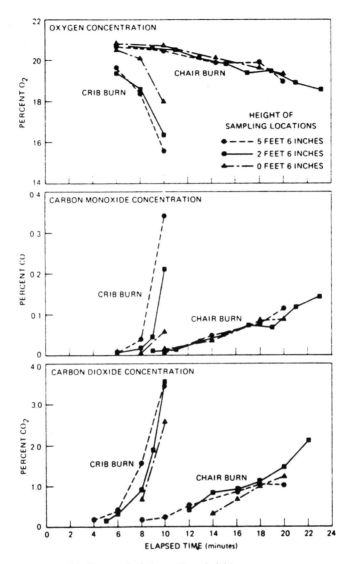

(*a*) Oxygen depletion at three heights.
(*b*) Carbon monoxide concentrations at three heights.
(*c*) Carbon dioxide concentrations at two heights.

FIG. 5—*Fire atmosphere analyses in a second-story bedroom* [16].

up during a one-story building burn [*17*] (Fig. 6). The objective of the study was to establish the available safe evacuation time during a fire and the hazard of CO and $CO_2$. Analysis was by on-site gas chromatography.

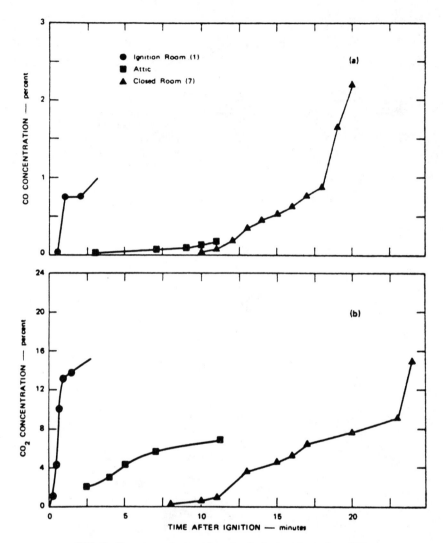

FIG. 6—*Fire atmosphere analyses in single-story house burn* [17].

## Fire Atmosphere Analysis of Corridor Fire Burns

Analyses were made of the buildup of fire gases during carpet fires [*15*] (Fig. 7). The objective of this study was to assess and identify the relative hazard of toxic gases in the combustion of various types of carpets. A corridor facility was used for these measurements. Compositions were measured on site using gas chromatography.

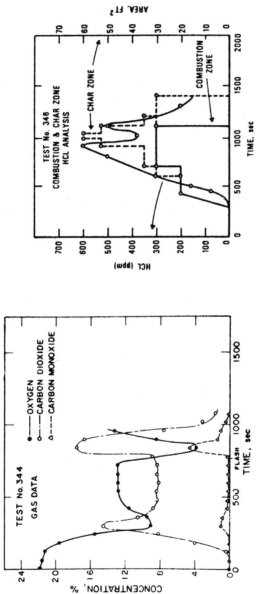

FIG. 7—*Fire atmosphere analysis in corridor carpet fire [15].*

## Flame Structure

One area of combustion research which has received considerable attention is the analysis of steady-state flames [8,18]. Here, microprobing is used in conjunction with mass spectrometry or gas chromatography or both. In some cases, molecular beam inlet probes [1,5] have been used so that the concentrations of radicals, atoms, and ions could be determined. For one-dimensional flame systems, a satisfactory quantitative interpretation can be made which allows the determination of fluxes and rates of production or disappearance of the various species. In less symmetric cases, lateral diffusion and heat conduction must be considered. Formal analysis has not been made, although it would appear feasible with existing computers. A matrix of measurements would be necessary for the multidimensional cases.

In the example of Fig. 8, measurements are given for a burning polymethyl methacrylate rod in air [19].

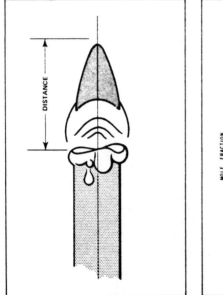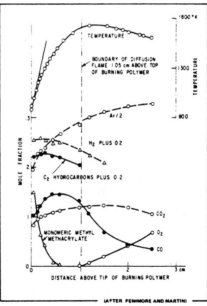

FIG. 8—*Structure of polymethyl methacrylate flame burning in the candle mode* [19].

## Idealized Fire Atmosphere Characterization

As can be seen by the previous examples, the analysis of fire atmospheres has not approached either the sophistication, which would be required for a complete quantitative understanding, or the state of the art of sampling and analysis.

Figure 9 suggests what might be accomplished with present technology. The objective would be to sample a fire atmosphere in sufficient detail to assess the hazard to life and property. A probe could be designed to sample volumes of 1 m$^3$ every 10 s. In the burn of a house with a volume of 300 m$^3$, a reliable picture could probably be assembled statistically by instrumenting 10 percent of these volumes. This would require at least 30 probes and associated analytical equipment, computer control, and data analysis.

The important points about the sampling system are: (a) the probe would systematically traverse through the volume being characterized, (b) the sampling rate and probe size would be adjusted to simulate normal nasal breathing, (c) initial separations of the sample would be made so that separate analyses could be made of components which appear in the nose, tracheal bronchial tree, and the lung, (d) analyses would only be made for known hazardous components, and (e) sampling could be automated and computer controlled.

Such a system would characterize a 1-h fire with 400 000 measurements. This would be an extremely detailed and informative description of a fire, but many types of fires would need study before reliable conclusions could be reached, and, as mentioned in the beginning of this paper, this leaves the small-scale and flame structure information still to be determined. Each of these is a job of at least the same order of magnitude.

## Summary

The good news is that a usable, relatively complete fire characterization is within the capabilities of present technology, and the costs of some of the keys to implementation, that is, computer-controlled instrumentation and data handling, are rapidly declining. The bad news is that it will be some time before the resources can be gathered to carry this out, and much remains to be done to make intelligent use of the information which might be gathered.

## References

[1] Fristrom, R., "Flame Sampling for Mass Spectrometry," *International Journal of Mass Spectra and Ion Physics,* Vol. 14, No. 15, 1974.
[2] Petajan, J., Voorhees, K., Packham, S., Baldwin, R., Einhorn, I., Grunnet, M., Dinger, B., and Birkey, M., "Extreme Toxicity from Combustion Products of a Fire Retarded Polyurethane Foam," *Science,* Vol. 187, No. 743, 1975.
[3] Autian, J., "Toxicologic Aspects of Flammability and Combustion of Polymeric Materials," University of Utah Conference, College of Engineering, Salt Lake City, Utah, 1970.
[4] Wagner, J., "Survey of Toxic Species Evolved in Pyrolysis and Combustion of Polymers," *Fire Research Abstracts and Reviews,* Vol. 14, No. 1, 1972.
[5] Fristrom, R., "Probe Measurements in Laminar Combustion Systems," Project Squid Workshop on Combustion Measurements, Purdue University, West Lafayette, Ind., 1975. *Fire Research Abstract and Reviews,* Vol. 16, 1974.

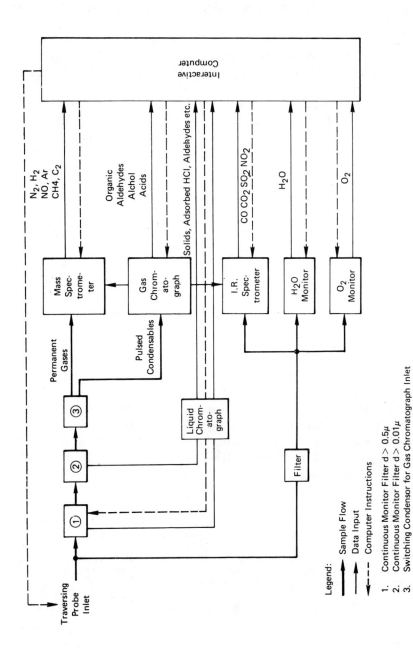

FIG. 9—*Idealized fire atmosphere sampling system.*

[6] Bilger, R., "Probe Measurements in Turbulent Combustion," Report PURDU-CL-75-2, the Combustion Laboratory, School of Mechanical Engineering, Purdue University, West Lafayette, Ind., 1975.

[7] Sawyer, R., "Experimental Studies of Chemical Processes in a Model Gas Turbine Combustor," *Emissions,* W. Cornelius and W. Agnes, Ed., Plenum Press, New York, 1972.

[8] Fristrom, R. and Westenberg, A., *Flame Structure,* McGraw Hill, New York, 1965.

[9] Tinè, G., *Gas Sampling and Chemical Analysis,* Pergamon Press, New York, 1961.

[10] Croce, P. A., "A Method For Improved Measurement of Gas Concentration Histories in Rapidly Developing Fires," FMRC #21011.3 Factory Mutual Corp., Norwood, Mass., 1975.

[11] Birkey, M., private communication.

[12] Liebman, S. A., Ahlstrom, D., and Sanders, C., "Automatic Concentrator Gas Chromatographic System," *American Laboratory,* Vol. 7, No. 9, 1975.

[13] Larsen, D., Rony, P., and Titus, J., "The Structure of a Microcomputer," *American Laboratory,* Vol. 7, No. 10, 1975.

[14] Zinn, B., "Combustion Products from Building Fires," *Proceedings,* National Science Foundation/Research Applied to National Needs Conference on Fire Research at Georgia Institute of Technology, School of Aerospace Engineering, Atlanta, Ga., 1974.

[15] Birkey, M., "Physiological and Toxicological Effects of the Products of Thermal Decomposition of Polymeric Materials," NBS Special Publication #411, National Bureau of Standards, Washington, D.C., 1974.

[16] Robison, M., Wagner, P., Fristrom, R., and Schulz, A., "The Accumulation of Gases on an Upper Floor During Fire Build-up," *Fire Technology,* Vol. 8, No. 278, 1972.

[17] Alger, R., "A Mobile Field Laboratory for Fires of Opportunity," NOLTR 73-87 Naval Ordinance Laboratory, White Oak, Silver Spring, Md., 1973.

[18] Fenimore, C., *Chemistry in Premixed Flames,* Pergamon Press, New York, 1964.

[19] Fenimore, C. and Martin, F., "Burning of Polymers," *Mechanisms of Pyrolysis, Oxodation and Burning of Organic Materials,* L. Wall, Ed., 4th Materials Research Symposium, National Bureau of Standards, Washington, D.C., 1970.

# DISCUSSION

*A. F. Robertson*[1] (*oral discussion*)—Bob, I think that this is a very relevant discussion. In consideration of gas analysis during accidental fires, it's very common to think of placement of a probe at one point in the room and make your analysis and report these were the gases and concentrations that were observed during this experiment. But I think that so many of us forget the fact that this was a single probe at a single location in a particular type of fire, so that I would like to recommend that all of us who are concerned with these types of problems remember the limitations on the types of gas analysis that we are doing today. Perhaps you have suggestions or comments on the concern that should be expressed for this limitation.

*R. M. Fristrom* (*oral discussion*)—Well, I would share your worries. To mitigate this problem, I have suggested that one should have a probe

[1] Senior scientist, Center for Fire Research, Institute for Applied Technology, National Bureau of Standards, Washington, D.C. 20235.

which traverses the volume it is supposed to characterize (in other words, that a probe which moves from point to point and develops a spatial average which may give a more meaningful characterization of the system than one taken at a single point). A complete characterization would require a rake of probes that completely blanketed the situation. One is not interested in the microturbulence aspect, but more in the question of variations in the volumes that a person takes in a breath. Such an average would really be relevant for life safety. So I am also suggesting that perhaps one should design probes not for chemistry but to simulate the nose intake.

*J. R. Gaskill*[2] (*oral discussion*)—Bob, isn't it still important that you have the question of sorption/desorption behavior of the gases on the various particular fractions as they travel from the probe to the analyzer?

*R. M. Fristrom* (*author's oral closure*)—This is a serious problem and one which must be addressed. There are ways of minimizing it. If you can locate your analytical device immediately adjacent to your sampling point, you are in great shape. But I can't afford to burn up my mass spectrometer every time I want to make a measurement. Continuous flow teflon-lined sample tubes are one possibility.

*C. M. Huggett*[3] (*oral discussion*)—Bob, you have outlined a very nice scheme for accumulating data and I would agree that, given a sufficient number of millions of dollars, we can get a pretty accurate picture of what goes on in a fire. Have you given any thought to what you are going to do with that data when you get it, and how many millions it's going to take to be able to interpret it in some useful fashion? I am particularly concerned with the problem of being able to draw conclusions concerning the physiological effects of the hundreds of compounds that you're going to find in your fire gas, possible interactions between these compounds, and interactions with such factors as temperature, oxygen concentration, and the like.

*R. M. Fristrom* (*author's oral closure*)—The only thing I can say is that I would hope by the time that we accumulate the data there will have been a great deal of thought given to how it is interpreted. You are quite right: if someone doesn't sit down and think about the information and try to draw conclusions, it is completely useless.

[2]Fire protection engineer, Lawrence Livermore Laboratory, University of California, Livermore, Calif. 94550.
[3]Chief program chemist, Center for Fire Research, Institute for Applied Technology, National Bureau of Standards, Washington, D.C. 20235.

*J. H. Petajan,*[1] *R. C. Baldwin,*[1] *R. F. Rose,*[1] *and R. B. Jeppsen*[1]

# Assessment of the Relative Toxicity of Materials: Concept of a Limiting Toxicant

**REFERENCE:** Petajan, J. H., Baldwin, R. C., Rose, R. F., and Jeppsen, R. B., "Assessment of the Relative Toxicity of Materials: Concept of a Limiting Toxicant," *Fire Standards and Safety, ASTM STP 614,* A. F. Robertson, Ed., American Society for Testing and Materials, 1977, pp. 285-299.

**ABSTRACT:** Experiments were conducted on male Long-Evans rats instrumented for measurement of vital functions and a conditioned avoidance response. An intra-arterial cannula was used for removal of blood samples. Rats were exposed to combustion products of three polymeric materials. A National Bureau of Standards smoke chamber and a smaller "static" chamber were used for exposures. Material A produced a syndrome of carbon monoxide (CO)-induced anoxia, the severity of which depended only upon the amount of material degraded and not upon the mode of combustion (heat flux, flaming, or nonflaming). Material B produced a syndrome of epilepsy and carboxyhemoglobin levels below 10 percent. Material C produced a metabolic acidosis and mild CO-induced anoxia, the severity of which was related to the amount of material degraded, irrespective of the combustion mode. The combustion products of Materials B and C produced intoxication syndromes distinctly different from the syndrome of CO-induced anoxia produced by Material A. "Limiting" toxicants or substances with high biological activity may be present in combustion products and produce unique intoxication syndromes.

**KEY WORDS:** fires, rats, vital statistics, conditioned responses, combustion products, toxicants, carbon monoxide, anoxia, epilepsy, acidosis, intoxication syndromes

Intoxication syndromes produced on exposure to products of degradation of many materials in our environment constitute a serious hazard to human life [1-4].[2] So many materials are involved that detailed study of each one is not possible. Because conditions, such as heat flux, the physical

---

[1] Professor, Department of Neurology, College of Medicine; assistant professor, Department of Material Science, College of Engineering; and technologists, respectively, Flammability Research Center, University of Utah, Salt Lake City, Utah 84132.
[2] The italic numbers in brackets refer to the list of references appended to this paper.

configuration of the material, flaming or nonflaming modes of combustion, systems for conducting smoke, etc., may account for variation in degradation products, procedures for analysis of the toxicological hazard for even a single material can be complex and time consuming [5]. Therefore, it seems reasonable to attempt to identify those materials which produce unusual or exceptional intoxication and to approach the problem of coping with such materials from several different directions. For example, a material producing unique intoxication may be modified so that toxicity is reduced. For some materials, the possibility of human injury may be so negligible that toxicity is not at issue. Structural or flammability characteristics may be so superior that some materials must remain in use despite significant toxicity. Many factors must be considered before a material is accepted for general use. Segregation of materials on the basis of exceptional toxicity is an important first step in the logic leading to decisions concerning usage. Criteria must be established which set limits for intoxication. This paper suggests a method for defining exceptional toxicity. In addition, a premise is put forward which states that exceptional toxicity may be based upon the presence of a single substance among the combustion products which has a high degree of biological activity capable of producing the essential features of the intoxication syndrome. This toxicant, sometimes acting in concert with other substances present in the mixture of combustion products, is responsible for a specific intoxication effect, an alteration in the function of some organ or system which cannot be explained on the basis of anoxic anoxia or carbon monoxide (CO)-induced anoxia alone. Removal of this toxicant from the mixture should eliminate the effect. Acting alone, the toxicant may not be capable of producing the specific intoxication effect owing to the adjuvant action of other substances present in the mixture.

## Method for Assessment of Exceptional Toxicity

Detection of exceptional toxicity requires an accurate assessment of the state of anoxia. Previous experiments have utilized sling-restrained Long-Evans rats instrumented for measurement of vital functions (electrocardiogram (EKG), electroencaphalogram (EEG), respirations, and special functions such as peripheral nerve conduction velocity, and cortical-evoked response). Through an intra-arterial cannula, blood samples were obtained for determination of carboxyhemoglobin, oxyhemoglobin, $pO_2$, serum bicarbonate, pH, total carbon dioxide ($CO_2$), and base excess during the course of intoxication. A conditioned avoidance response requiring the rat to maintain left hind-limb flexion to avoid a shock was also used. Control data for changes in vital function and blood parameters and behavioral alterations occurring during exposure to CO, have been obtained and reported previously [6–8].

The following is a summary of the data obtained during exposure of rats to CO. The conditioned avoidance response was lost when carboxyhemoglobin reached 49 ± 2 percent during exposure to 2500-ppm CO. At this level of intoxication, a fire victim would be unable to help himself. Except for the degree of CO-induced anoxia present, vital functions were only slightly perturbed: slight decreases in respiratory rate, heart rate, and blood pressure were observed. These changes were completely reversible with the return of carboxyhemoglobin to control levels. Carrying the exposure further to produce carboxyhemoglobin levels between 60 and 70 percent produced a steady decline in heart rate, respiratory rate, and blood pressure until a state of ischemic anoxia was reached, at which point, the EEG became isoelectric, cortical evoked response was lost, and the ventral caudal nerve failed to conduct. At this point, immediate resuscitation is required if death is to be prevented. This level of intoxication establishes the limits within which it is possible for others to save a fire victim.

Following the exposure, and depending upon the level of intoxication reached, various degrees of morbidity may ensue, ranging from impaired learning ability to demyelination of the central or peripheral nervous system [9–12]. These effects may require as long as two weeks to occur in the rat. In man, months or longer are required.

The evaluation of CO-induced anoxia in the rat has focussed upon three levels or kinds of intoxication: (a) loss of the survival response (the victim is unable to save himself); (b) loss of resuscibility (the victim must be saved by others); and (c) morbidity (there are pathological effects following exposure). Carboxyhemoglobin levels associated with these levels of intoxication have been determined for ambient CO concentrations in the moderate range. Very high CO concentrations may cause given levels of intoxication to occur at slightly lower carboxyhemoglobin levels. For this reason, ambient CO concentration should be known when experiments are conducted to assess animal response to combustion products. The ability to load and unload CO is an indication of the adequacy of cardiopulmonary function, normal loading being dependent upon adequate lung ventilation and circulation of blood. Such factors as pulmonary edema (fluid in the lungs), bronchial irritation, or breath-holding may delay the uptake of CO and modify the rate of increase of carboxyhemoglobin.

During screening experiments on animals not fully instrumented for physiological and behavioral assessment, a clinical examination has been applied which has been roughly correlated with the results of more detailed measurements [13]. This assessment was conducted in a 3 by 4-ft enclosure on freely moving animals following exposure to combustion products. The battery of tests which was routinely performed is summarized as follows.

*Exploratory Behavior*—The animal is dropped from a height of 30 cm into the center of a 25-cm-diameter circle. The time to place one extremity, excluding the tail, outside the circle is determined.

*Nuzzle Response*—The response to stimulation of the nose and vibrissae with a cotton swab is rated on a 0 to 4 + scale, representing responses ranging from indifference to an attack on the swab, 2 + being normal.

*Lid, Corneal, and Ear Reflexes*—This is determined with a wisp of cotton and rated on a −/+ scale.

*Pain Response*—This is determined by the number of pin pricks rapidly applied to the dorsum of the hind foot which are required to produce withdrawal. A scale of 0 to 4 + is used, 2 + being normal.

*Righting Reflex*—This is determined by dropping the animal upside down from a height of 30 cm. The rat normally can land on all four feet in the time required. A scale of 0 to 4 + is used, 4 + being normal. Failure to place an extremity properly accounts for one lower grade.

*Posture*—This is graded on a 1 to 5 + scale and expressed as a fraction of normal.

*General Observations*—Respiratory distress, lack of spontaneous activity, abnormal movements, etc. are observed.

This protocol is most appropriately used as an initial screening procedure. Systemic effects can often be detected which permit prediction of toxicants that may be affecting the test animal. Exploratory and nuzzling responses and righting reflexes rapidly disappear and are lost at approximately the 50 percent carboxyhemoglobin level. Reflexes depending upon brain-stem integrity, such as the lid, corneal, and ear reflexes, and gross effects upon posture, occur as the animal reaches the 60 to 70 percent carboxyhemoglobin level, which produces ischemic shock. Arterial blood pressure falls below that sufficient to maintain oxygenation of vital organs.

## Concept of the Intoxication Syndrome

The physiological, behavioral, and biochemical changes which accompany exposure to CO constitute a syndrome or group of responses which consistently occur in the anoxic state. This syndrome has been used as a reference against which syndromes produced by other toxicant mixtures can be compared and contrasted. The purpose of this comparison is to detect those toxicants which produce their effects by means other than anoxia. We have assumed, for the purposes of this comparison, that a comparable degree of anoxic anoxia will result in the same physiological and behavioral end points as those produced by CO-induced anoxia, an assumption that requires further investigation and verification.

When either CO-induced anoxia or anoxic anoxia exist, evaluation of the physiological and behavioral response, pathological evaluation of the animal sacrificed postexposure, and knowledge of ambient CO concentration can usually provide an explanation for the degree of anoxia present. In the experiments to be described, blood parameter evaluation, clinical examination, and an examination of general physiological status have been found adequate for discrimination of the syndromes which are not produced by a state of anoxia.

## Exposure Chambers

Two kinds of chambers have been used in these experiments. The first is a modified National Bureau of Standards (NBS) smoke chamber which has been used for exposure of four Long-Evans rats suspended radially in a sling apparatus, so that all shared the same breathing zone. One animal contained an intra-arterial cannula, so that blood samples could be obtained during and following exposures while behavioral and physiological observations were being made on the remaining animals. This chamber, until recently, did not permit detailed physiological monitoring and has been used for "before and after" experiments. This "screening" procedure has been used for the detection of intoxication syndromes produced when materials were combusted at varying heat fluxes. Monitoring of smoke density, CO, and other gases was customarily done.

A second chamber, a 40-litre acrylic "static box," provides facilities for complete physiological, behavioral, and biochemical monitoring of a single animal. The amount of material combusted was scaled down appropriately from that used in the NBS chamber. The course of the intoxication process was followed, and vital functions were monitored. Behavior was sampled continuously throughout the exposure. The conditioned avoidance response just described was found to be the most convenient measure of behavior.

## Conditions of Material Degradation

Sample mass, heat flux (orientation and configuration), duration of sample heating, and the presence of a pilot flame constitute the critical variables of the degradation process. Sample mass was selected to produce a moderate degree of intoxication such as the loss of the conditioned avoidance response. Loss of resuscibility (shock) sometimes occurred when the intoxication syndrome became severe. At times, sample mass had to be adjusted up or down to produce the desired degree of intoxication. Similarly, heat flux was changed to achieve a desired level of combustion in order to produce a given intoxication end point. In general, samples were combusted over a range of heat fluxes to determine whether or not alterations in the intoxication syndrome occurred. Samples were heated for 10 min (NBS chamber) or 5 min (static box). The total animal exposure time was 20 min. Three or more heat fluxes were used for some materials. The use of different heat fluxes was compelled by the observation that the composition of combustion products may change significantly as a consequence of flux, thereby altering the intoxication syndrome.

## Materials

Three polymeric materials were studied; for purposes of this discussion, they will be labeled A, B, and C. The materials were selected because they produce states of intoxication which demonstrate both an anoxic syndrome and specific syndromes easily differentiated from anoxia.

## Results

### Material A

Seventeen experiments were performed with one male 250 to 300-g Long-Evans rat in each experiment using the static chamber. Results are presented in Table 1. Fluxes of 2.5, 5.0, and 7.5 w/cm² were used to combust the material in both flaming and nonflaming modes. At 5 and 7.5 W/cm², spontaneous ignition occurred. The syndrome produced on exposure to the combustion products was consistently one of CO-induced anoxia. Serum pH, bicarbonate, and excess base were not significantly affected until intoxication at shock levels was produced. Behavioral responses were those predicted from the degree of CO-induced anoxia present. Carboxyhemoglobin levels present when the conditioned avoidance response was lost were only slightly less than 50 percent. Increasing the mass of material combusted increased the level of carboxyhemoglobin and the severity of the anoxia syndrome (Fig. 1). Varying heat fluxes or changing the combustion mode did not produce any qualitative change in the nature of the intoxication syndrome.

Examination of the remaining data in Table 1 reveals that the decrease in oxyhemoglobin can be accounted for by the elevation in carboxyhemoglobin. The levels of $pO_2$ and $pCO_2$ indicate that pulmonary ventilation was adequate except when intoxication was severe enough to produce anoxic shock. This is predictable from the fact that the loading of CO approaches that seen during exposures to the pure gas. It is essential to know that hemoglobin and hematocrit (the percentage by volume of red blood cells in the blood) is normal, because rats with decreased oxygen-carrying capacity will not withstand the same elevation in carboxyhemoglobin levels as will normal animals. Elevation in white blood cell count (WBC) may accompany the stress of exposure. This would not seem to be a predominant effect. It is necessary to obtain all these values in order to detect other causes for anoxia, such as impairment of oxygen transport at

TABLE 1—*Material A.*

| Animal | Time | White Blood Count | Hematocrit | Oxyhemoglobin | Carboxyhemoglobin | Hemoglobin | pH | $pCO_2$ | Base meq/L Excess | $HCO_3^-$ | $pO_2$ | $O_2$ Saturation | W/cm² | Mass (initial) | Mass (final) | Comments |
|---|---|---|---|---|---|---|---|---|---|---|---|---|---|---|---|---|
| 58 | 0 | 10.0 | 49.0 | 0 | 0 | 16.4 | 7.48 | 41.7 | 6.1 | 30.9 | 82.0 | 86.5 | 2.5 | 0.57/2.1 | | NF[a] |
|  | 19 | 19.1 | 49.1 | 90.8 | 7.4 | 15.9 | 7.51 | 22.0 | −2.0 | 17.6 | 105 | 88.2 | | | | |
| 59 | 0 | 23.3 | 41.9 | 84.6 | 0 | 15.0 | 7.44 | 38.6 | 2.0 | 26.0 | 74 | 95 | | | | |
|  | 24 | 15.9 | 33.3 | 85.7 | 11.8 | 11.7 | 7.49 | 29.5 | 1.2 | 22.4 | 72 | 95 | | | | |
| 62 | 0 | 11.0 | 39.6 | 93.4 | 0 | 14.2 | 7.46 | 29.3 | −1.1 | 20.7 | 74.0 | 95.4 | | | | |
|  | 17 | 15.7 | 41.3 | 87.6 | 7.5 | 15.7 | 7.55 | 18.6 | −1.8 | 16.4 | 73.0 | 96.3 | | | | |
| 64 | 0 | 9.4 | 38.3 | 84.4 | 26.5 | 13.4 | 7.51 | 35.7 | 5.1 | 28.5 | 72.0 | 95.5 | | | | |
|  | 20 | 29.3 | 37.4 | 62.2 | 36.6 | 13.2 | 7.55 | 25.4 | 1 | 22.3 | 70.0 | 95.9 | | | | |
| 63 | 0 | 17.6 | 35.2 | 98.2 | 0 | 12.3 | 7.47 | 16.9 | −8.0 | 12.2 | 76.0 | 96.0 | 2.5 | 0.53/2.05 | | F[b] |
|  | 22 | 15.5 | 28.2 | 83.0 | 14.0 | 11.4 | 7.47 | 26.4 | −2.2 | 19.1 | 70.0 | 94.9 | | | | |
| 60 | 0 | 20.8 | 36.3 | 92.7 | 0 | 12.7 | 7.47 | 24.8 | −3.2 | 17.9 | 81.0 | 96.5 | | | | |
|  | 22 | 8.5 | 33.2 | 47.4 | 40.9 | 11.6 | 7.48 | 28.9 | −1.0 | 21.5 | 85.0 | 96.9 | | | | |
| 66 | 0 | 12.7 | 39.5 | 89.2 | 11.0 | 14.3 | 7.47 | 29.2 | −1.5 | 21.1 | 75.0 | 95.7 | | | | |
|  | 27 | 11.2 | 36.1 | 62.6 | 41.5 | 13.3 | 7.49 | 35.2 | 3.5 | 26.8 | 65 | 93.9 | | | | |
| 69 | 0 | 11.10 | 42.5 | 95.0 | 2.8 | 13.1 | 7.44 | 25.1 | −4.8 | 16.9 | 64.3 | 93.9 | | | | |
|  | 22 | 13.3 | 27.6 | 68.0 | 32 | 10.4 | 7.45 | 21.8 | −6.5 | 15.0 | 60.8 | 94.0 | | | | |
| 70 | 0 | 16.5 | 44.0 | 94.9 | 3.9 | 15.4 | 7.48 | 28.4 | −1 | 21.1 | 73 | 95.4 | | | | |
|  | 19 | 13.2 | 28.3 | 71.5 | 30.8 | 12.3 | 7.49 | 21.2 | −6 | 20.7 | 63 | 93.5 | | | | |
| 53 | 0 | 14.8 | 44.9 | 95.6 | 10 | 15.8 | 7.50 | 30.8 | 2.7 | 24.0 | 79.0 | 96.4 | 5 | 1.1/2.05 | | F at 37 s |
|  | 25 | 22.8 | 44.8 | 42.6 | 51.6 | 15.2 | 7.38 | 32.1 | −4.8 | 18.7 | 75.0 | 94.6 | | | | |
| 54 | 0 | 12.5 | 42.4 | 86.2 | 5.0 | ... | ... | ... | ... | ... | ... | ... | | | | |
|  | 21 | 12.0 | 37.5 | 45.3 | 48 | ... | ... | 1 | ... | ... | ... | ... | | | | |
| 55 | 0 | 9.6 | 47.6 | 95.2 | 0 | 9.8 | 7.52 | 20.6 | −3.8 | 16.9 | 67.0 | 95.1 | | | | |
|  | 20 | 12.0 | 27.1 | 53.2 | 39.4 | 14.8 | 7.46 | 52.7 | 12 | 37.3 | 59.0 | 90.6 | | | | |
| 56 | 0 | 8.0 | 45.9 | 88.1 | −1.0 | 13.3 | 7.24 | 34.6 | −12.1 | 14.2 | 67.0 | 89.5 | | | | |
|  | 22 | 9.7 | 37.2 | 64.7 | 37.5 | 12.3 | 7.46 | 29.1 | −1.4 | 20.6 | 78.2 | 95.4 | | | | |
| 71 | 0 | 16.0 | 34.4 | 89.4 | 3.00 | 13.0 | ... | ... | ... | ... | ... | ... | 7.5 | 1.65/2.04 | | F at 5 s |
|  | 28 | 16.5 | 37.2 | 36.9 | 68.2 | 9.6 | 7.02 | 39.3 | −20.4 | 9.7 | 22.1 | 17.0 | | | | |
| 72 | 0 | 10.8 | 27.5 | 88.8 | 9.4 | 15.1 | 7.36 | 31.6 | −6.2 | 17.5 | 68.0 | 92.5 | | | | |
|  | 18 | ... | 44.1 | 40.3 | 54.3 | 14.7 | 7.16 | 49.5 | −11.8 | 17.0 | 58.0 | 81.0 | | | | |
| 73 | 0 | ... | 43.1 | 90.5 | 0.9 | ... | 7.50 | 35.1 | 5.3 | 27.3 | 65.0 | 93.9 | | | | |
|  | 16 | ... | ... | 32.7 | 62 | ... | ... | ... | ... | ... | ... | ... | | | | |
| 74 | 0 | 18.3 | 38.3 | 96.6 | 0 | 12.6 | 7.53 | 26.8 | 1.9 | 22.4 | 63.0 | 94.1 | | | | |
|  | 20 | 14.4 | 34.4 | 31.3 | 62.5 | 11.4 | 7.09 | 43.6 | −17.9 | 12.7 | 74.0 | 88.1 | | | | |

[a] NF = nonflaming.
[b] F = flaming.

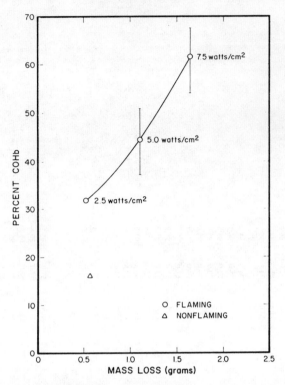

FIG. 1—*Percent carboxyhemoglobin following a 20-min exposure versus mass of Material A lost during combustion.*

the cellular level, loss of oxygen-carrying capacity by red cell hemolysis, etc.

## Material B

Three experiments were performed, one each at 2.5, 5.0, and 7.5 W/cm$^2$. There were four 250 to 300-g male Long-Evans rats in each experiment. Partway through the exposure, animals were noted to convulse. All animals convulsed following removal from the chamber. At higher heat fluxes, and with a greater percentage of material combusted, most of the animals died. All behavioral responses were depressed as a consequence of the convulsions, but the startle reflex was enhanced prior to the development of seizures. Automatic, purposeful, but inappropriate, behavior was also seen. In static box experiments, spike and polyspike activity was noted in the electroencephalogram. Carboxyhemoglobin values were near control levels, and oxyhemoglobin levels were nearly normal (Table 2). At all three fluxes, this material produced a syndrome of epilepsy, grand mal

TABLE 2—*Material B.*

| Number of Animals | COHb | Time to Move From 25 cm Circle | Response to Pain (pin prick) | Other Responses |
|---|---|---|---|---|
| 12 rats | 5 to 6% | >60 s (<6 s = normal) | none | myoclonic jerks, status epilepticus, death |

seizures, and minor motor seizures. The syndrome may be clearly distinguished from that of an anoxic state.

## Material C

Twenty-three experiments were performed with one 250 to 300-g male Long-Evans rat in each experiment. Exposures were conducted in the static box during nonflaming combustion at 1.0 (2 rats), 1.5 (8 rats) and 2.5 (13 rats) W/cm². Lower heat fluxes were required for this material because of the severity of the intoxication syndrome produced at 5.0 W/cm² or above. As can be seen from Figs. 2 and 3, the syndrome con-

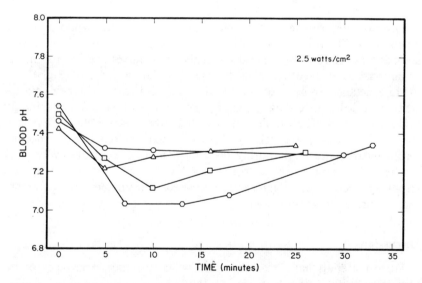

FIG. 2—*Arterial blood pH for individual rats versus time during exposure to the combustion products of Material C.*

sists of acidosis associated with a mild to moderate increase in carboxyhemoglobin. Serum bicarbonate and base excess were significantly decreased, the findings being typical of metabolic acidosis. (Increases in car-

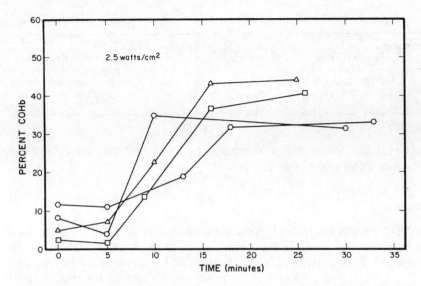

FIG. 3—*Arterial blood percent carboxyhemoglobin for individual rats versus time during exposure to the combustion products of Material C.*

boxyhemoglobin of the order found here do not produce acidosis of this magnitude). The acidosis was present at all heat fluxes used. More severe acidosis was found in rats exposed at higher heat fluxes and greater masses of material combusted, but considerable variability was found (Fig. 4).

More detailed analysis of the pH decrease and carboxyhemoglobin increase revealed that, during the initial few minutes of exposure, respiratory acidosis occurred. The pH decrease resulted from breath holding. Evidence for this was seen in the delayed rise of carboxyhemoglobin and normal serum bicarbonate levels during the first 5 min of exposure. When pH reached its lowest levels, a rise in carboxyhemoglobin occurred, and serum bicarbonate was significantly decreased. Measurement of respiratory rate confirmed this impression.

## Discussion

Three materials have been discussed which produce syndromes of intoxication quite distinct from one another.

The first material produces a syndrome of CO-induced anoxia which does not vary qualitatively over a range of heat fluxes; the severity of the syndrome depends upon the amount of materials degraded. The second material produces a syndrome of epilepsy which is quite easily differentiated from the syndrome of CO-induced anoxia. The third material produces a syndrome of metabolic acidosis and moderate CO-induced anoxia, the

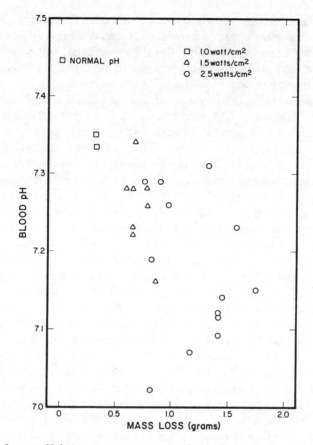

FIG. 4—*Lowest pH during exposure versus mass lost in the combustion of Material C.*

severity of which depends upon the intensity of the exposure which, in general, is a function of the amount of material degraded.

Despite the fact that the mode of combustion and heat flux produced qualitative alterations in the combustion products, no significant qualitative change in the syndrome was seen for any of the materials studied. One can conclude from this that the essential factor(s) responsible for the syndrome are not affected by the mode of combustion or heat flux. The condition may be unique for these materials, and results may be quite different for other materials.

In the case of Material B, a specific systemic effect—the production of abnormal excitability of the central nervous system—constitutes the limiting effect of the combustion products. This unusual effect comprises the primary toxicant action of the combustion products. Identification of the toxicant responsible for this effect, determination of factors which decrease or enhance the effect, and an investigation of the mechanism of action of the toxicant must then be carried out. The complexity of such investigations and the considerations of material usage may be such as not to warrant further investigation. On the other hand, some toxicants may produce effects of considerable interest to biological scientists, and investigations may be warranted on this basis alone.

The third material investigated acidifies the blood and produces a mild to moderate degree of CO-induced anoxia. The effect produced is similar to that which would occur if an acid were injected directly into the blood. Questions concerning the nature of the acidifying process, its possible cause by a single agent, and the influence of other factors which may modify the acidifying process remain to be answered. As already stated, the syndrome produced is distinct from that of anoxia alone.

A toxicant which may be primarily responsible for a specific effect such as seizures or acidosis could exert its action because of the complementary action of other toxicants in the mixture. The ability of a primary or limiting toxicant to enter the blood may depend upon such factors as the depression of protective pulmonary reflexes, modification of cell membrane permeability, or the operation of certain carrier mechanisms such as solvents or particles. In order to test the hypothesis that such adjuvant factors are important to the production of the intoxication syndrome by a single toxicant, use can be made of synthetic gas mixtures. With such mixtures, it may be possible to reproduce the intoxication syndrome. The syndrome may or may not be reproduced by a single toxicant. It is important to emphasize that the limiting toxicant may exert its effect only in the presence of other substances present in the mixture of combustion products. It is possible that the effect would disappear if the toxicant were removed from the mixture, even though the effect was not produced by the toxicant administered alone. In order to assess completely any intoxication syndrome, experiments must be performed to differentiate and characterize limiting toxicant criteria with these factors in mind.

Accurate characterization of the intoxication syndrome depends upon the creation of reproducible ambient conditions. As yet, the methods used for combusting materials leave much to be desired. If flaming occurs partway through the combustion process, then the intoxication syndrome may be different from that produced in either the flaming or nonflaming mode. Also, it is difficult to produce uniform decomposition of the material. All of these factors will affect the intoxication syndrome produced. Much work remains to be done in this area of research.

It is possible to rank the concerns for fire safety: the first concern is for flammability itself; second, if ignition occurs, it is desirable that the intoxication syndrome produced on inhalation or in contact with combustion products be reversible. The physician responsible for treating the fire victim should not be confronted with a unique and complicated problem. Materials which produce disabling syndromes when relatively small amounts of material are degraded must be identified and their use modified accordingly. It is essential to describe the intoxication syndrome produced and to determine what factors modify its severity; it is important to know whether the syndrome changes qualitatively when the conditions of material decomposition are changed. These are not impossible goals to achieve, and the resultant data can serve as a rational basis for the assessment of relative toxicity of materials.

## Acknowledgments

The authors wish to acknowledge the assistance of I. N. Einhorn, J. H. Futrell, K. J. Voorhees, R. W. Mickelson, and Dr. S. C. Packham for their many scientific contributions and for assisting us in this research program. We wish to further acknowledge J. B. McCandless and T. L. Blank who provided technical assistance throughout this study. Acknowledgement is also made for the assistance and advice given by Dr. M. M. Birky, Acting Chief, Program for Toxicology of Combustion Materials, National Bureau of Standards, who served as Project Monitor for NBS Contract No. 5-9006 which supported this research. Partial support was also received from the National Aeronautics and Space Agency and the National Science Foundation (Research Applied to National Needs) Program.

## References

[1] Drinker, C. K., *Carbon Monoxide Asphyxia,* Oxford University Press, New York, 1938, p. 138.

[2] Dufour, R. E., *Bulletin of Research,* Underwriters' Laboratories, Inc., Northbrook, Ill., Vol. 53, July 1963.

[3] Zikria, B. A., Sturner, W. A., Ostarjeon, N. K., Fox, C. L., and Ferrer, J. M., *Pathophysiology and Therapy Annals,* New York Academy of Science, Vol. 150, 1968, p. 618.

[4] Kishitani, K., *Journal of Faculty of Engineering,* University of Tokyo, Tokyo, Japan, Vol. XXXI, No. 1, 1971.

[5] Gaskill, J. R., *Journal of Fire and Flammability,* Vol. 1, July 1970, pp. 183–216.

[6] Petajan, J. H., Packham, S. J., Dinger, B. G., and Frens, D. B., "Effect of Carbon Monoxide on the Nervous System of the Rat," *Archives of Neurolology,* in press, 1975.

[7] Grunnet, M. and Petajan, J. H., "CO-Induced Neuropathy in the Rat: Ultra Structural Changes," *Archives of Neurology,* in press, 1975.

[8] Packham, S. J., Petajan, J. H., Christiansen, E., and Dinger, B. G., "An Animal Model for Combustion Toxicology," *Journal of Fire and Flammability,* in press, 1975.

[9] Ginsberg, M. D. and Myers, R. E., *Transactions of the American Neurology Association,* Vol. 97, 1972, pp. 207–211.

[10] Dutro, F. R., *American Journal of Clinical Pathology,* Vol. 22, 1952, pp. 925–935.

[11] Snyder, R. D., *Neurology,* Vol. 20, 1970, pp. 117–180

[12] Blackwood, W., McMenemy, W. H., Meyer, A., Normal, R. M., and Russell, D. S., *Greenfield's Neuropathology,* Williams and Wilkins Co., 1966, pp. 240–244.

[13] Petajan, J. H., Voorhees, K. J., Packham, S. C., Baldwin, R. C., Einhorn, I. N., Grunnet, M., Dinger, B. G., and Birky, M. M., *Science,* Vol. 187, 1975, pp. 742–744.

# DISCUSSION

*M. M. Birky*[1] (*oral discussion*)—You mentioned, in the beginning of your talk, the inadequacy of doing carboxyhemoglobin measurements on human fire fatalities. On the studies that have been reported, there are other measurements being done, such as the cardiovascular examination and pathology of the respiratory tract. Would you suggest other measurements besides those being done or in place of those measurements?

*J. H. Petajan* (*author's oral closure*)—I am sure that a complete postmortem examination is being done, and, of course, we recognize that postmortem values of acid base balance and oxyhemoglobin aren't going to be of much value, but I think we should recognize that because a victim's carboxyhemoglobin may be in the range of 40 to 50 percent does not necessarily mean that the cause of his death was entirely due to carbon monoxide. I think you also must realize that, if you do detailed analysis of the coronary vascular system, coronary narrowing by itself may not be the cause of death. The pathologist can usually give you a pretty good idea. He can tell you whether there has been a recent thrombosis, that is, a plugging off of the vessel, or whether the muscle of the heart shows evidence of recent absence of blood supply. My point in mentioning this was that I don't want people here to simply go away with the impression that because 50 percent or so of fire victims have a carboxyhemoglobin in the 40 to 50 percent range, which we know is disabling, this is *ipso facto* evidence that we have proof that carbon monoxide is totally responsible for the death of these individuals. I don't think the investigators would assert that either. And the identification of intoxication syndromes of other kinds, in addition to the one of CO-induced anoxia, makes it important for us to examine the fire situation and try to get what information we can about materials that were burned there and, perhaps, to get some blood samples from fire victims. We may find that syndromes found in animals represent problems at the fire scene. We have anecdotal evidence that some of these syndromes are being produced during the acute inhalation of smoke.

*A. F. Robertson*[2] (*oral discussion*)—I wonder to what extent there is work in progress to relate these kinds of experiments to the real thing that we are interested in, that is, the disabling effects on humans?

*J. H. Petajan* (*author's oral closure*)—Well, if you refer to the carboxyhemoglobin levels, that is, as a starter, the carboxyhemoglobin levels

[1] Chief, Section of Combustion Toxicology, National Bureau of Standards, Washington, D. C. 20234.

[2] Senior scientist, Center for Fire Research, Institute for Applied Technology, National Bureau of Standards, Washington, D. C. 20235.

which produce a loss of an avoidance response, incapacitation, or shock in a rat, they are very comparable to those for man. It is simply a matter of the rate at which these levels are achieved. The primary emphasis in our laboratory has been to develop an animal model system for evaluating the relative toxicity of materials so that we can try to quantify this toxicity. We are trying to come up with a single factor or set of factors which we can use to describe a material. Components we are considering include the heat flux required to produce a given intoxication end point, the amount of material that is degraded to produce this end point, and the dosage difference between that required to produce 50 percent animal mortality (LD-50) and that which results in a loss of function in 50 percent of the animals (a type of safety factor). These various components may hopefully be combined to yield a measure of material toxicity. This is about as far as we have gone.

*A. F. Robertson (oral discussion)*—Are we safe in (I realize that we have to work with models and this is an animal model) feeling confident that this is a really valid predictor of most of the effects of the gases on man?

*J. H. Petajan (author's oral closure)*—Well, I could say that observations on the rabbit, which has a more suitable pulmonary system for evaluation of inhalent intoxication, have given similar results. Speaking as a biologist, the rat is a mammal. We are talking about mammalian systems which differ quantitatively, but probably do not differ qualitatively, with respect to responses we might see in man. Now the ability to extrapolate to man requires a knowledge of differences in metabolic rate, heart rate, respiratory rate, and differences in the amount of dead space in the lung. The rat is not a good animal for studying the direct effect of inhalants on the bronchi, etc. But I think that experiments or studies must be done to examine the possibility that different materials may produce different kinds of pulmonary syndromes or other effects in humans. These should preferably be studied at the fire scene or very soon thereafter. At the present time, clinicians do not differentiate smoke inhalation injuries of different types. They are gradually becoming aware that poly (vinyl) chloride (PVC), for example, can produce chronic lung problems and more morbidity than some of the other kinds of smoke. I think we need more studies of human intoxication.

*I. A. Benjamin*[1]

# Development of a Room Fire Test

**REFERENCE:** Benjamin, I. A., **"Development of a Room Fire Test,"** *Fire Standards and Safety, ASTM STP 614,* A. F. Robertson, Ed., American Society for Testing and Materials, 1977, pp. 300–311.

**ABSTRACTS:** This paper has attempted to indicate the need for full-scale room and compartment tests at this time, indicating two possible uses of the compartment test (for approval, or for validating small-scale tests) and has discussed a few of the factors which affect the design of the tests, using the choice of ignition source as an example of the problems involved in the test design. Reference is made to the recommended practice developed by the task group of American Society for Testing and Materials Committee E-39.10.01.

**KEY WORDS:** fires, fire tests, fire safety

Compartment fire tests have been conducted with increasing frequency in recent years without any standardization of procedures. There is a need for an agreed-upon procedure for conducting tests in rooms, so that results can be understood and interpreted by those people who must evaluate them.

The need has grown because of a belief on the part of many people in the fire safety community that the existing small-scale tests are not adequate to evaluate hazard in a consistent manner for all materials. The discrepancy between simulated real-life situations and small-scale tests has been noted by several experimenters. The problem was probably brought to a head when the Federal Trade Commission (FTC) issued their proposed rule on cellular plastic products [1].[2] The impact of the rule is best summarized in the following paragraph.

It is an unfair method of competition and an unfair or deceptive act of practice—to refer directly or indirectly to any standard or certification—unless such standard certification or representation is substantiated by a reasonable basis, which shall include competent scientific tests which determine, evaluate, predict or describe said combustion characteristic under actual fire conditions

---

[1] Chief, Fire Safety Engineering Division, Center for Fire Research, National Bureau of Standards, Washington, D. C. 20234.

[2] The italic numbers in brackets refer to the list of references appended to this paper.

to the extent that such tests are recognized as probative given the then existing state of the art.

In Note 3 of the same section, it also states:

> The Commission considers that neither the standard entitled "Standard Method of Test for Surface Burning Characteristics of Building Materials," variously denoted as ASTM E 84, UL 723, NFPA 255,—alone constitutes a "reasonable basis—under actual fire conditions" as required by this section.

As a result of the FTC action, there has been an intensified search within both industry and the building code community to determine a reasonable basis for substantiating performance of materials. The end result of this search has more or less centered on the concept of using a room test for the evaluation of wall and ceiling linings in lieu of, or as a supplement to, the existing E 84 test procedure. American Society for Testing and Materials (ASTM) Committee E-39, in response to the need for a standard room test established a task group to develop a recommended practice.

The room test concept has also been adopted by International Conference of Building Officials. In Section 1717c of a recent amendment of their building code, they state that plastic foam may be specifically approved, based on approved diversified tests such as, but not limited to, tunnel tests, fire tests related to actual end use such as a corner test, and an ignition temperature test. As a result of this change, the research committee of the Uniform Building Code has been attempting to develop a standardized corner test to be used in conjunction with the application of their code.

A room test can serve for many different purposes. Two common objectives are:

1. As an acceptance test to qualify products or constructions for use as wall and ceiling linings.

2. To establish the validity of various small-scale tests, which, in turn, may be used in lieu of the large-scale room test as acceptance tests.

The large-scale or room test is supposed to simulate the real world; but what is the "real world"? what is the "real fire"? The real world represents a range of probabilities. For example, the probability of a given fire load in a room will vary with the type of occupancy. However, even within a given occupancy, the probability of a given fire load has a wide range. In an office building, we can have fire loads from as low as 1 psf in a conference room to over 25 psf in a storage room. Knowing the existence of this possible range of fire loads in a building, one has to choose a reasonable fire load to determine what the scenario of a room test for product acceptance should be. By a scenario, we mean a postulated probable occurrence of ignition and the subsequent chain of fire growth. In addition, the scenario should have a reasonably probable relationship of that occurrence to the nature of the occupancy. For example, if one takes the most extreme case for an office building, then all offices should be evaluated for over 25 psf of fire loading in each compartment. However, this is not a probable level, and, therefore, a lower loading should be used.

The design of the large-scale test to accept products therefore requires the definition of a scenario related to their actual end use and one which reflects the probability of the incident.

In the work at the National Bureau of Standards (NBS), we have given considerable thought to developing scenarios, since the room and compartment tests that we conduct are predicated upon scenarios. An example of the importance of the scenario can be illustrated by the fact that the significance of the hazard of a particular combination of wall and ceiling linings in a room can be made to reflect the fire load that the scenario postulates. One can choose a fire load of such magnitude that its contribution will mask the effect of the linings. With this scenario, one can show from a large-scale test that the wall linings are insignificant, and, therefore, one should not be concerned about them. However, one can also choose another scenario in which the fire load is very light, and the only way for the fire to spread from one item to another is via the wall linings. With this scenario, the linings become most important. We have considered this latter case to be a more typical scenario. Under this scenario, the wall lining becomes a critical factor in fire spread and hazard, since flashover will not usually occur in a sparsely furnished room if the wall lining does not contribute.

A good illustration of the effect of choice of scenario is a comparison of the work done by Bruce [2] and Ferris [3] in the 1950's. Bruce conducted tests at Forest Products Laboratories in an 8 by 12 ft test room with an 8-ft-high ceiling. The scenario called for a "typical room" with furniture which gave a fire loading of 3.9 psf. Most of this weight was concentrated in the middle of the room, and the fire was ignited in the middle of the room. The closest point of the room furniture to the wall was 18 in. Under this test arrangement, Bruce concluded that:

> The nature of the walls, whether plaster, fiber insulation board, or plywood, had little or no effect on the time or temperature of the critical point and only a small effect on the flashover. In fact, the buildup of fire in the room was such that the flames and hot gases from the chair and table burning in the center of the room went up towards the ceiling, and the ceiling caught on fire at approximately the same time as the walls.

On the other hand, Ferris used a different scenario in his work at the Australian Commonwealth Experiment Station when he was developing a flame spread test for wall linings. This work was based on experience with a full-scale room test. In starting the study, Ferris looked at various types of furniture fires in a 12 by 13 ft room, 9 ft high. He found that

> Certain fires, particularly those involving kerosine heaters, or those in which plywood furniture was ideally placed, were so intense that any linings would add little to the initial overall hazard; other fires involving certain bedding and chairs were so feeble as to have little effect on any lining boards.

As a result of these full-scale tests, his scenario, designed to specifically evaluate the wall linings, was an arbitrary theoretical fire of medium intensity within the observed range of possibilities. The fire was simulated with a gas flame and was sized to best illustrate the effect of linings on the fire growth in the room.

These two examples illustrate how, by using scenarios with different ignition levels to represent the real world, two investigators could arrive at different conclusions on the hazard of wall linings.

The other key item in the design of the room test for acceptance testing is the criterion to be used. The criterion depends upon the nature and type and definition of the hazard that one wishes to guard against. The ASTM E-39.10.01 task group, in their draft, Recommended Practice for Room Fire Tests [4], discuss some of the criteria that might be used.

With our present state of knowledge, the criteria most normally applied deal with the rate and growth of fire and the time to flashover. One usable criterion is the elapsed time from ignition to flashover in the room. There is no currently accepted standard for this time; a 10-min criterion would probably be a socially acceptable guide for safety, based on the performance expectation that has been existing for many past years in the normal residential type of furnishings and construction environment. For more critical occupancies, this criterion could conceivably be lowered.

Flashover has been found, in a recent series of tests [5], to correlate with an average air temperature in the range of about 550 to 600 °C at the 5-ft level in the center of the room. A more direct indication of flashover might be the radiant energy flux incident on the floor. A level of about 2 $W/cm^2$ at the center of the floor is suggested as an equivalent to the incipient flashover in the room. The most convincing indication of flashover can come from the ignition of cellulosic indicator panels. Consequently, one criterion that may be employed in a room fire test would be the time to flashover, as measured by the thermocouples in the upper part of the room by a radiometer or total heat flux gage at the center of the floor or by the ignition of indicator panels.

The potential for growth and spread of fire within the room of fire origin is another criterion which is useful for determining the ease with which a major fire may develop. This criterion may be applied by evaluating the size of the igniting source, in terms of energy release, whether from a wood crib, waste basket, or a gas burner, needed to cause flashover in the room or flame emergence from the doorway of the room. This criterion, however, requires a series of tests to determine the critical size of ignition source required.

Another criterion that has been applied is an evaluation of the fire-damaged area on the wall and ceiling for a fixed size of ignition source (for example, a 20-lb wood crib) for the specimen under test. The rate of fire growth can also be documented photographically and used as an adjunct criterion.

The choice of criterion for acceptance may qualify some products under one criterion and disqualify them under another criterion. This is very interestingly illustrated in the recent publication by Underwriters' Laboratories on "Flammability Studies on Cellular Plastics and other Building Materials Used for Interior Finishes" [6]. The report uses, as a criterion for acceptance, the determination of whether the room reached flashover. This is evaluated by determining if a temperature of 649°C is reached at the ceiling or if full ceiling involvement occurred. The report concludes that: "the flame spread classification of materials developed in the standard 25-ft tunnel test corresponds with the performance of those materials in corner and vertical wall full scale building geometry tests."

TABLE 1—*Time to full ceiling involvement (with 20-lb crib).*

| Sample | Flame Spread Classification | Time | Material |
|--------|------------------------------|------|----------|
| CORNER TESTS | | | |
| J | 54 | 10:10 | untreated fiberboard |
| Q | 59 | 1:22 | foam plastic |
| AD | 75 | 0:40 | foam plastic |
| AE | 100 | 8:45 | red oak |
| ROOM TESTS | | | |
| S | 22 | 1:20 | foam plastic |
| A | 23 | 1:40 | foam plastic |
| G | 23 | N[a] | treated plywood |

[a]Not reached in 20 min.

If, however, one wishes to use a different criterion for hazard (for example, the time to flashover in the room), then the conclusion just cited is no longer valid. Looking at Table 1, constructed from data in the report, one can see that, for materials with the same flame spread classification in either a corner or a room test, the time to flashover varies by almost an order of magnitude. This is a significantly different interpretation of the effectiveness of the ASTM E 84 tunnel test in determining the hazard of materials since it shows that cellulosic materials used as wall linings would produce a different degree of hazard than foamed plastic materials, even though both have the same flame spread classification. From this discussion, we can see that the definition of the hazard from wall linings and the criteria for acceptance is one of the most difficult problems facing us today, but one that must be faced if we are to evaluate materials and make progress in the field of materials technology.

The second use for the room or compartment test is to validate the significance and meaning of the many small scale-tests that are being or have been proposed for material evaluation. This use is not new and has been attempted many times in the past, generally with the indication that the existing test methods are not always well validated by the full-scale or room tests. This problem has been summed up in a recent paper by Malhotra, in which he states:

> The early tests were devised with only a limited understanding of laws governing fire growth and severity and, while in the last decade the knowledge has improved, the material technology has progressed at a tremendously fast rate with demands for more and more fire tests outstripping the gains in knowledge. It is therefore not surprising that the ratio of satisfactory to unsatisfactory tests has deteriorated.

Some of the work attempting to validate small-scale tests is of interest here. For example, Tourrette [7] from Centre Scientifique et Technique du Batiment reported on a series of room tests which were designed to evaluate wall and ceiling linings. In this report, the values from the "epiradiateur" tests, the French standard, were compared with the results of the large-scale tests. He concluded "one consequence which we find from the study of the charts is the poor correlation between the reaction-to-fire classification and the periods proceeding general conflagration." In other words, the small-scale test did not predict time to flashover.

Christian and Waterman [8], in a series of studies for the Society of the Plastics Industry, used a corridor to evaluate the meaning of the flame spread classification from the ASTM E 84 test method, using test materials on the wall and the ceiling. The fire was originated in a 10 by 15-ft burn room with a 8-ft-high ceiling. The room was attached to the corridor and the flames and hot gas that left the room went into the corridor. They looked at the times required for the fire to travel the length of the corridor with various lining materials. Their conclusion was that "it is clear that placement of the materials in the order of ascending tunnel test flame spread ratings does not quite place them in the order of increasing flame spread rate or decreasing time in the full-scale corridor."

Ferris [3] in the work mentioned earlier, compared the results from the horizontal spread-of-flame test (BS 476: 1953) with time for the flame to spread up the wall to the 9-ft level. He noted "the coefficient of correlation is 0.69, which is just significant at the 2 percent level, showing that there is a trend, although the relationship is not particularly good."

Ferris did find a good correlation between his large-scale room tests and the vertical spread-of-flame test he later developed. This was to be expected since the test was developed to duplicate the observed performances of wall linings in the large-scale test room.

Hird and Fischl [9] of the Joint Fire Research Organization conducted a series of tests, using a 12 by 18 by 9-ft-high room. They compared their results with the existing standard, BS 476. Although there was general agreement between their full-scale tests and the standard, they found it significant that: "the highest classification of the surface spread of flame test includes boards with a wide range of performance, and, where they are to be used as linings for both walls and ceiling of compartments containing an appreciable amount of combustible material, the surface spread of flame classification is not a sufficient indication of the fire hazard."

Work done by Fang [5] at NBS using a 10 by 10-ft room, 8 ft high, showed a correlation between the temperatures developed in the upper part of the room and the E 84 flame spread classification. However, the work was limited to cellulosic and other conventional linings and did not include any plastic foam boards.

The just-mentioned selected reports indicate that existing small-scale tests do not always correlate with the full-scale simulation, and, until we have small-scale tests for materials which validly indicate the hazard for a given scenario, we must be limited by room or compartment tests to properly evaluate wall and ceiling linings.

Task Group 4 of ASTM E-39.10.01 has developed a recommended practice for room fire tests [4] to provide a guide to the research or testing agency. The purpose of the recommended practice is to provide a guide which compiles the information and experiences now available from those actively engaged in this type of testing. The task group hesitated at attempting to develop a test method, since the state of the art is still in development. However, the document that was produced does attempt to give, in detail, the various factors which should be taken into account in the design of a room or compartment test.

The task group discusses such things as a compartment design, the nature and type of specimen, ignition sources, instrumentation, and safety practices in their recommended practice. Each of these represents an important range of choice which must reflect, to some degree, nature and type of scenario. For example, just looking at the decision to be made on the size, nature, and type of the ignition source indicates the complexity involved in deciding on the nature of a room test.

The ignition source in any fire will have a large effect on the total performance in the room. In fact, the ignition source may become a critical factor in determining how and when a lining may be involved in a fire. In some of the earliest work on evaluating linings, Corson and Lucas [10] did an extensive study of ignition sources, preparatory to conducting a series of full-scale room tests to evaluate linings and wall constructions. The tests were conducted in a 14 by 20-ft room, 12 ft high. A series of wood cribs were used of 5, 7½, 10, 20, and 30 lb weight. The heat release of the fire was analyzed, and the temperatures developed in the room and the maximum flame height were recorded. For a room of this height, it took a 20-lb crib to have the flames hit the ceiling.

Whether the flames hit the ceiling during a test will determine the relative importance of the combustibility of the wall and ceiling linings. If the flames do not touch the ceiling, then the combustibility of the wall linings determine if the ceiling material will become involved, since the fire can only reach the ceiling by traveling up the wall lining. If, however, the flames reach the ceiling because of the size of the ignition source, then the combustibility of the wall linings will be a lesser or negligible factor in determining if ignition and spread occurs on the ceiling. To illustrate this point, in work done by Fang [5], it was found that the importance of the wall in relation to the ceiling lining depended upon the size of the igniting fire. For an igniting fire that did not touch the ceiling—simulating a small upholstered chair—the combustibility of the wall lining became a major factor in the time to flashover in the room. However, when a larger simulated upholstered chair, which had flames that impinged directly on the ceiling was used, then the effect of the wall lining became far less important, and the combustibility of the ceiling lining governed the time to flashover. In this case, the choice of the size of chair can lead one to differing conclusions on the importance of the wall versus the ceiling lining.

In a recent series of tests conducted by Underwriters' Laboratories [6], the effect of the size of the crib was again studied. Use was made of a "corner test," which is a two-sided room having 8 by 8-ft walls and a section of ceiling at the intersection of the side walls, with the other two faces left exposed. A series of ignition sources were investigated, using 2 or 4 lb of combustibles in waste containers or a 20 or 30-lb wood crib. Temperatures at the ceiling were recorded for the various fire loads. From this work, they developed a qualitative comparison of the ignition energy necessary to cause the corner assembly to flashover. They then related this to the flame spread rating of the lining under the E-84 test method. This work illustrated, in a very useful manner, the fact that the size of the ignition source in a corner or room test using wall and ceiling linings will have a major effect on whether or not the linings become involved and whether flashover will occur.

The scenario for the large-scale or real-life fire must therefore include some postulate regarding the nature and type of the ignition source. The ignition source has been varied in our work at NBS, depending upon the scenario chosen. For example, in an evaluation of the flammability of seats on buses and subway trains we have used a 2-oz paper bag of trash and a 2-lb folded newspaper, since, by consensus, these scenarios were agreed upon as representing ignition sources likely to be found in such transit vehicles. On the other hand, when studying wall linings in rooms, we have used a small or large upholstered chair as the item to be ignited, depending upon the height of flame we wish to investigate. This scenario represents a fire in a piece of furniture in a room corner and is used to evaluate the potential of the wall lining to spread the fire from the igniting source.

## Summary

This paper has attempted to indicate the reason for the need for full-scale room and compartment tests at this time. I have indicated the two possible uses of the compartment test (for approval or for validating small-scale tests) and have discussed a few of the factors which affect the design of the tests, using the choice of ignition source as an example of the problems involved in the test design. Reference is made to the recommended practice developed by the task group of ASTM Committee E-39.10.01.

## References

[1] *Federal Register,* Vol. 40, No. 142, 23 July 1975, p. 30842.
[2] Bruce, H. D., "Experimental Dwelling—Room Fires, Forest Products," Laboratory Report No. 1941, April 1959.
[3] Ferris, J. E., "Fire Hazards of Combustible Wallboards, Commonwealth Experimental Building Station," Special Report No. 18; Oct. 1955.
[4] Report of Task Group 4, ASTM Committee E-39.10.01, Recommended Practice for Room Fire Tests (draft for Letter Ballot).
[5] Fang, J. B., "Fire Buildup in a Room and the Role of Interior Finish Materials," NBS Technical Note 879, National Bureau of Standards, June 1975.
[6] Castino, G. T., Beyreis, J. R., and Metes, W. S., "Flammability Studies of Cellular Plastics and Other Building Materials Used for Interior Finish," File Subject 723, Underwriters' Laboratories Inc.
[7] Tourette, J. C., "L'Influence des Revetements sur le Development de L'Incendie," CSTB Report No. 144, Centre Scientifique Et Technique Du Batiment, Paris.
[8] Christian, W. J. and Waterman, T. E., "Fire Behavior of Interior Finish Materials," *Fire Technology,* Aug. 1970.
[9] Hird, D. and Fischl, C. F., "Fire Hazard of Internal Linings," National Building Studies, Special Report No. 22, Joint Fire Research Organization, 1954.
[10] Corson, R. C. and Lucas, W. R., "Life Hazard of Interior Finishes," Laboratory Report No. 11760, Factory Mutual Laboratories, June 1950.

# DISCUSSION

---

*G. T. Castino[1] (oral discussion)*—Let me just raise one point. The comparison that you have made is based on the tunnel calculation as it is now used, and that is the way, of course, we reported the data in the report. Actually, we have found in subsequent studies that if one compares the tunnel numbers with the data developed in the room burns on the same basis, in other words, on the basis of time rather than distance, that the correlations are quite a bit better.

*I. A. Benjamin (author's oral closure)*—There have been several attempts made to change the classification scheme in the E-84 test method. It was done originally about five or seven years ago, I think, and it is currently under consideration again. And I suspect that this new classification system may be an improvement.

*G. T. Castino (written discussion)*—There are some problems created for

---

[1] Managing engineer, Underwriter's Laboratories, Northbrook, Ill. 60062.

us by the statements in this paper, as well as the related data in Table 1. It is agreed that, once sufficient radiant flux is developed at the ceiling elements of the corner test geometry, the difference between the time of involvement of cellulosic materials and cellular plastic materials is unreasonably large. However, the basis of the flame spread calculation in the tunnel is primarily by distance of spread. Therefore, Underwriter's Laboratories (UL) used area of fire involvement as the basis of comparison. The author has used time as the basis. In separate studies, carried out at UL, it appears that, when compared on the same basis, the tunnel "performance" does correspond to the "performance of the same lining materials over a range of ignition source severities in corridor, corner, and vertical wall geometry enclosures.

Further, the author elected to use the ceiling involvement time, which is perfectly acceptable with the exception that it is the longest time (last to occur) because it requires full ceiling area coverage in flaming. If one were to use the time to reach the end of the wall-ceiling intersection (the area or zone of maximized radiation reinforcement) an entirely different picture unfolds; this, of course, assumes that one is interested in equivalently relating these measurements in the tunnel and the corner tests on the basis of time. Using area of involvement as UL did, the overall relationships that led to the finding quoted by the author is valid.

Another point to be considered is that the author selected materials as given by Table 1 in his paper, exhibiting flame spread classifications by the tunnel test, in excess of 50 to compare with corner fire tests using 20-lb wood crib fire sources. If one deals with materials with flame spread classifications of less than 50, once again, one finds that the UL finding is shown to be more creditable.

*I. A. Benjamin (author's written closure)*—The key point in the paper is that the evaluation of hazard depends upon the criteria chosen. The paper suggests that one valid criterion for hazard can be expressed as the time it takes to produce flashover in a room. Using this criterion, then, the tunnel test is not an accurate predictor of the hazard, as evidenced by Table 1. Mr. Castino is correct in saying that one should not expect the tunnel to give valid results, based on this particular criterion. I agree with this statement and was trying to point out that the tunnel is limited in the nature and type of fire behavior which it is equipped to predict.

*H. L. Malhotra[2] (written discussion)*—The compartment fire test which appears to be gaining a great deal of popularity and status is a large-size fire test which reproduces the use of materials or products on which information is required in a very realistic manner. There are, however, many other variables related to the design of the igniting source, the range of

---

[2] Head, Building and Structures Division, Fire Research Station, Borehamwood, Herts, United Kingdom.

qualitative observation, and the quantitative measurements that are possible to make which necessitate the introduction of deliberate simplification. Consequently, the full-scale fire test deals only with one of the many possible situations and is usable to give direct answers, like many of the other existing tests, to the whole range of problems. The need will continue to exist for exercising caution in the application of data, and its interpretation will require a great deal of expert knowledge which does not fully exist today.

*H. Paul Julien*[3] (*written discussion*)—Mr. Benjamin has presented an interesting paper on the prospects for a compartment (corner or room) fire test. He makes the point (on page 304 and in Table 1) that all materials with the same flame spread classification (FSC) in the ASTM E 84 tunnel test do not necessarily exhibit the same time to flashover (or full ceiling involvement) in a corner or room test. I can agree fully with this conclusion.

Table 1 of this paper was constructed from data in a comprehensive report of Underwriters Laboratories Inc. (Ref 6). Unfortunately, the entries were selected in such a way as to convey the erroneous impression that cellulosic products produce a lower "degree of hazard" than do foamed plastic materials of the same FSC.

Thus, the limited data shown on Corner Tests in Table 1 were chosen to indicate a lack of correlation for four samples of intermediate FSC. But the UL study (Table 21) also showed that a 20-lb crib did not cause full ceiling involvement of any cellular plastic or cellulosic material with an FSC of 28 or less (a total of 15 samples). Also, all 8 samples with an FSC above 100 had a time to flashover of less than 5 min. On an overall basis, then, there is clearly a considerable degree of correlation.

A glaring omission occurs under Room Tests in Table 1. The Sample R from Table 23 of the UL report should be included to present accurately the results of that work and to support Mr. Benjamin's basic point about a lack of correlation. Sample R is a "foam plastic" with 27 FSC which did not reach full ceiling involvement in 20 min. I propose the addition of one line under Room Tests in Table 1, as follows:

|     |   | Room Tests |        |                 |
|-----|---|------------|--------|-----------------|
|     | S | 22         | 1:20   | Foam plastic    |
|     | A | 23         | 1:40   | Foam plastic    |
|     | G | 23         | N[a]   | Treated plywood |
| Add | R | 27         | N[a]   | Foam plastic    |

[a] Not reached in 20 min.

With this addition, Table 1 now illustrates that materials with the same

[3] Manager, Advanced Technology and Testing, Jim Walter Research Corp., St. Petersburg, Fla. 33702.

FSC and of the same generic group can show a vastly different behavior in the full-scale room test.

I believe this change is necessary to avoid a gross distortion of this part of the UL study. Table 23 clearly shows that Sample R (cellular plastic) outperformed Sample G (treated plywood). Sample G sustained "nearly full fire involvement" (page 60 of Ref 6), but Sample R did not. Also, the room lining area affected by fire involvement was 356 ft$^2$ (89.5 percent) for Sample G versus only 70 ft$^2$ (17.6 percent) for Sample R. Furthermore, the Sample S included in Table 1 is not a building material at all, but is a cellular plastic specimen specially mutilated for experimental purposes (page 9 of Ref 6). Thus, Mr. Benjamin's condensation of the UL data on room tests appears to this observer to be something less than impartial.

*I. A. Benjamin (author's written closure)*—It was not our intention in discussing the data in the UL report to focus attention on particular products or types of products. The conclusion drawn from these data, with which Dr. Julien concurs, is that a given flame spread classification (for example, FSC = 25) in the ASTM Method E 84 tunnel test is not a valid indicator of whether or not flashover will occur in a room fire test.

Sample R consisted of a sheet of cellular plastic with an exposed foil facing, the presence of which had an appreciable effect on fire performance. For this reason, it could not be compared with foams or board that were directly exposed to fire. Sample S was the identical cellular plastic sheet from which the foil face had been removed.

# Panel Discussion

*F. K. Willenbrock*[1]—Here is how I suggest that the panel operate.

Since this is an unusually capable panel, we will pose three complex questions and give them about 3 min to respond. The first question is, "How far can fire standards take us?" The second question is, "How much fire safety is enough?" The third is, "What are the biggest challenges in improving fire safety?" We have asked each of the panelists to consider these questions. They have been given the choice of answering all three in 3 min or answering one question in depth for 3 min. You can see that we have a lot of confidence in the ability of the panel to respond to this unusual intellectual challenge.

We will work in the following manner: I will ask each panel member who wishes to make a statement to limit himself to approximately 3 min in length. The other panelists will then ask questions of an informative nature. Questions of an argumentative nature should be saved for later. After all of the panelists have had an opportunity to speak, we will then throw the floor open for questions from the audience. I should like to use the same ground rules for the audience: no question can take more than 3 min. In addition, I have to be able to understand the question or it is not a question.

With that introduction, I would now like to turn to our panelists. We have decided to start with Mr. Burgun. Mr. Burgun is an architect. He has been very active in fire safety considerations. He is the chairman of the National Fire Protection Association Life Safety Committee.

*J. A. Burgun*[2]—I would like to speak on all three of the questions and as briefly as possible. How far can fire safety standards take us; not very. You could have the best standard in the world, but, until you actually implement those printed pages into the reality of building design and construction, you have accomplished very little.

How much fire safety is enough? It is too much when it restricts and interferes with the quality of life style or freedom of society, and society is unwilling to accept the imposition. As an example, I could assure you that you would never die of a fire if I placed you in a plain concrete box, did not provide you with heat or light, allowed you no reading materials, smoking materials, or indeed even clothing. Your chances of dying in a

---

[1] Director, Institute for Applied Technology, National Bureau of Standards, Washington, D.C. 20234.
[2] Architect, Rogers, Butler and Burgun Architects, New York, N.Y. 10016.

fire would be very, very slim, but your quality of life would be pretty bad. There is a second point when there is too much fire safety, and that is when it consumes a disproportionate share of a given society's budget, and there are more pressing problems to be solved. As an example, we are working on a hospital project in Nigeria, and, believe me, we don't worry about fire safety. We have other problems far more pressing, such as the need for water and basic health care services.

Question 3: what is the biggest challenge in improving fire safety? As far as I'm concerned, it is finding an effective and reasonable way of retrofitting existing structures with good quality fire prevention measures.

*F. K. Willenbrock*—Thank you. Our next speaker is Larry Seigel. Larry is a research consultant for the U. S. Steel Research Center. He has been active for some time in fire research programs for his company as well as the industry in general.

*Larry Seigel*[3]—An advantage of being the second speaker is that you can immediately have a disagreement. I think I heard Mr. Bergun say that fire standards cannot take us very far, and the first sentence I wrote in my notes was, "Fire standards can take us as far as the public is willing to tolerate." But I don't think there is any real disagreement, because I suspect that is also what he has in mind. The cost of imposing fire standards will increase rapidly as the ultimate in fire safety is reached. And, of course, there are practical limits in any case because merely writing codes does not guarantee compliance with those codes. There have been fires that many of us may recall that illustrate that compliance did not occur, and, as a result, disasters did. But we must continue our efforts to improve fire safety; and to this end, I have heard many good suggestions made at this symposium. But the challenge that we face is how to implement many of the ideas within the framework of our codes and standards systems. Life safety is now becoming one of our major goals, and the products of combustion are now being recognized as a significant hazard. Methods for controlling the combustion products need to be developed, as has been suggested by one speaker at this conference. Perhaps the control of combustible materials at the manufacturing level may be one direction in which to go.

For thermal and structural performance of buildings, much has been said about engineering design methods, and these should be encouraged. But some method of acceptance of analytical design for fire endurance under building codes must be developed because such design methods are not generally recognized today. It may be necessary to invent new test procedures to establish basic performance of structural elements at elevated temperatures. Then, by new analytical methods, the complete performance of building assemblies may be subject to engineering design.

[3] Associate research consultant, U.S. Steel Research Center, Monroeville, Pa. 15146.

This procedure is not unlike that currently used in structural design where the strength characteristics of small specimens of material are determined by standard tests, and the structural performance of the complete assembly is then determined by calculation. Such designs are then certified by a structural engineer. We may be approaching the time when fire protection engineers will assume similar responsibilities.

*F. K. Willenbrock*—Thank you, Larry. Our next panelist is Wolfram Becker. Mr. Becker is a civil engineer. He is the head of the Fire Safety Technology group of Badische Anilin-und Soda-Fabrik industries in West Germany. He is vice chairman of the Deutsche Industrie-Normen Liaison Committee on fire tests and has been very active in both international and national standardization work. Mr. Becker.

*Wolfram Becker* [4]—First, I must say that I was a little bit in doubt if I should sit on the panel because of our different way of life in Germany, our different history, and language; I hope this won't cause misunderstanding.

Let me say at the outset that a fire is an event that threatens health and even life, and the possibility of its occurrence can never be excluded. The reason why fires can always happen is that, in our daily life, we are surrounded by sources of energy that we cannot do without. Fire protection will, therefore, provide only relative safety to men, even in the future, because we can't avoid these potential sources of ignition.

How far we are prepared to carry fire protection depends, on the one hand, on the need for safety of the individual, and, on the other, on the readiness of society to assume the burden of preventive measures. We cannot say today what hazard threshold will be acceptable to the individual and to society in the future; this applies as much to fire hazards as to any others. The expenditure and limitations required to attain a relative degree of safety are hard to estimate. They will depend greatly on the economic situation and on the presence of other threats to life, such as war and disease; they must also involve sacrifice of luxury to a certain extent.

In attempting to determine the future direction of fire protection, it is helpful to first take a glance into the past. At least until the turn of the century, fires presented a potential hazard to whole cities. The last great fires occurring in peacetime, in Hamburg, Chicago, and Toronto, are still remembered. Up to the end of the second World War, the aim of fire protection was clearly to restrict the extent of fire damage to a single building.

Nowadays, we attempt to contain fires within compartments of the building involved. The logical development in the future could be to aim

---

[4] Vice chairman, Deutsche Industrie-Normen, and engineer, Fire Safety Technology, Badische Anilin-und Soda Fabrik, Ludwigshafen, West Germany.

at preventing a fire from reaching the flashover stage. This tendency toward a more complex consideration can also be recognized in the development of standards, test methods for materials, and particularly building materials. Testing of the material alone and requirements referring only to the materials, which in the past were considered adequate, are being replaced by fire performance assessments of products in the situation in which they are to be used.

Nowadays, we have to consider the consequences of the situation, namely considerably more differentiated testing and investigations and critical questioning of conventional criteria. This trend will probably continue until the end of the century. We also have to get away from the idea of the single crucial experiment or test that can be regarded as the touchstone for prophecying fire risks.

In the future, the performance of a particular arrangement of products will be based on the results of a large number of tests involving various forms of thermal attack.

And now to answer the three questions: If we continue as we have begun, we should be in a position to make a real contribution to improving fire safety standards by adequately limiting hazardous situations and by relating experiment to practice.

Fire tests will, however, give no answer to the question of how much fire protection is needed. It might give an answer to the question how much fire protection is available. There will probably never be enough fire safety because human beings cannot stand the effects of even the smallest fire.

*F. K. Willenbrock*—Thank you, Mr. Becker. Any questions?

Our next speaker establishes a bit of the pattern we had in mind. We first asked Mr. Burgun, who is an architect, to speak. We then had two representatives of the industrial community who are concerned with supplying products which have fire safety aspects. We now turn to the research community, and our first speaker is Professor Witteveen. Professor Witteveen has been carrying on research in this general area. He is associated with the University of Tech Delft, as well as with Toegepast Natuurwetenschappelijk Onderzoek, and has been very active on the International Council for Building, Research, Studies, and Documentation and the International Standards Organization.

*J. Witteveen*[5]—I would like to speak on just one item: "What are the biggest challenges we face in fire safety today?" Perhaps you expect that I will suggest issues concerning scientific or technological problems. However, I don't think that the biggest challenges are in these areas; I feel

[5] Professor, Structural Mechanics, University of Tech Delft, deputy director, Institute Toegepast Natuurwetenschappelijk Onderzoek for Building Materials and Building Structures, Rijswijk, The Netherlands.

that the most important issues are problems of communication. Well, if you look at the partners who play important roles in fire safety, we have: the authorities, the designers, the fire services, and the fire scientists. All those partners have their own responsibilities and their own means to meet these responsibilities. I believe that, until now, the developments in the respective areas have taken place in isolation. As an example, research on improvements in firefighting has had little impact on research and standards on structural fire problems, and vice versa. Also the research output has had little effect on building regulations. In order to arrive at an effective fire protection situation, the developments should be interrelated. This means that the fire scientist must be aware of the needs and responsibilities of the other partners, and vice versa. This suggests that there is a need for communication in order to arrive at an integrated system. Recent tendencies, however, lead to a growing gap between the partners. They often seem to be talking completely different languages and focussing on their own limited responsibilities. I think breaking this trend is the major challenge in improving fire safety.

*F. K. Willenbrock*—Thank you. Any comments from the panel? If not, I will call our next speaker, Dr. John Lyons, Director of the Center for Fire Research at NBS.

*J. W. Lyons*[6]—The first question is how far can standards take us? I would agree with the other gentlemen who've said that, in and of themselves, they can take us nowhere. However, if standards are properly imbedded in building codes and in other regulatory measures, or, indeed, in the consciousness of our people, I think they can take us very far. The Center staff has, for the last year, in response to new Federal legislation, been looking at the question of how far technology can take us, given a reasonable level of technical effort in both the Federal and the private sectors. We estimate that, if we do those jobs that we are confident can be done in the research or engineering area, and if the results of this work are properly implemented by the rest of the community, a reduction of fire loss in this country of greater than 40 percent is entirely possible. This is just from the research component, and that leaves out the contributions of all the other things that one can think about that one might do about fire. So, I think standards can take us a long way if they're considered a part of this broader effort.

"How much is enough?" is a question that I think should properly be directed to the highest-level policy makers in our country. It would be presumptuous of me to try to answer that, but I will give the answer that the people's representatives in the United States have given. The Congress, in response to the National Commission's report, "America Burn-

[6] Director, Center for Fire Research, National Bureau of Standards, Washington, D.C. 20234.

ing," has certainly indicated its agreement that this country ought to try to cut its losses by half in a generation, and then we'll see whether we wish to cut them further. And finally, what is the greatest barrier? Let me address that strictly as a research question, because I think, if you broaden the question, you get into the issue of public apathy and indifference which is, perhaps, the biggest part of our difficulty. In the research area, I think Mr. Roux's speech yesterday highlighted the point better than I can. I think the biggest barrier is in providing the connecting links between, in, and among the various things we know how to measure or can learn how to measure, so that we can relate these things to the overall risk or potential for harm (as Hank Roux put it). Then, of course, if we can do that, I think we can devise the control measures to set or control that potential for harm at whatever point society wishes it.

*F. K. Willenbrock*—Thank you very much. Are there any questions the panel would like to raise? If not, we will now turn to our next speaker, Bob Blake. Bob is Director of Planning and Development for the Facilities Engineering and Property Management in the Department of Health, Education, and Welfare (HEW). Bob used to be a member of the National Bureau of Standard (NBS) Center for Building Technology.

*Robert Blake*[7]—Thank you, Karl. "How much fire safety is enough?" To sketch out a quick response to that question, or better, to rationally approach the question, we need a preface statement that shows the HEW accountability for fire losses, particularly loss of life in nursing homes. You are aware that a number of laws govern the Federal assistance to State and local agencies for new construction, alteration, mortgage insurance, etc., for medical facilities. In the case of MEDICAID and MEDICARE, the laws require a Federal compliance role to ensure life and fire safety of the patient. Pressures on HEW to assume an active role in this area in the early 1970's resulted in an HEW/NBS program to collaborate in the area of fire safety. It started in mid 1971 and is planned to continue until 1979. We can refer to the proposed NBS Flooring Radiant Panel Test as a specific example of program results.

It is now the clear objective of the work going on at NBS on behalf of HEW to specifically address the issues of facility cost, first cost and on-going cost, and the probable benefits in terms of life and property safety.

Public administrators, both elected and appointed, as well as the Civil Service, have not traditionally been pressed to explicitly and openly consider economic costs and benefits. Rather, an American ethic has prevailed that seems to say we can do what needs to be done if we will only generate the resources from our limitless supply to satisfy the need. Things are different now. Government at all levels seems to implicitly

[7] Director, Office of Planning and Development, Department of Health, Education, and Welfare, Washington, D.C. 20201.

understand that there are limits to growth, and that, when waste occurs, such waste of public capital funds is actually a waste of savings, and savings wasted are gone forever. If waste is too strong a word, then perhaps we could say less than useful expenditure of public capital funds. It is our clear expectation that the NBS Center for Fire Research will develop the tools that will make it possible for government facility programmers and the architect engineer to better articulate to the public at large the economic cost choices and the benefit alternatives in the area of life safety.

*F. K. Willenbrock*—Thank you, Bob. If there are no questions for Bob, we will turn to your next speaker, Joe Clark. Joe was also a member of NBS until a short time ago when he played a very key role in helping the new National Fire Prevention and Control Administration get underway as its first Acting Administrator and now as Associate Administrator for Fire Safety and Research.

*J. E. Clark[8]*—I think whatever remarks I might make would echo what Bob Blake has just said, which is that defining the alternative solutions to fire safety problems is one of our greatest challenges. Standards are certainly among those solutions. Presenting in an open forum the basis on which we choose from among those alternatives and noting that there are costs as well as benefits associated with whichever one we might choose are important. I don't think that there is much I can add, except to say that, in the area of nonregulatory solutions to fire safety problems, there are alternatives that have not been addressed very much by public policy makers. The questions of incentives, whether they be economic incentives or even psychological and motivational incentives, introduce other kinds of approaches that need to be investigated, along with regulatory and standardization approaches. Perhaps the Federal Trade Commission will have another view.

*F. K. Willenbrock*—Joe almost introduced our final speaker. No panel in Washington would be complete without having a lawyer aboard. So we are happy to have Eric Rubin here. Eric is from the Federal Trade Commission (FTC). He is concerned with compliance and consumer protection. He has been very active in the fire area through some recent FTC actions. Eric.

*Eric Rubin[9]*—My involvement is that of a lawyer, and I guess that speaks for my expertise by itself. I think, though our involvement has been in respect to polymeric materials, I should make it clear that what I am going to talk about goes for traditional as well as polymeric materials.

What is the greatest barrier to progress? Well, as a layman, when we

[8] Associate administrator, National Fire Prevention and Control Administration, U.S. Department of Commerce, Washington, D.C. 20230.

[9] Assistant director for compliance, Bureau of Consumer Protection, Federal Trade Commission, Washington, D.C. 20580.

first got involved in fire issues, we were absolutely appalled at the archaic state of the art with respect to the technological understandings of the phenomena itself. My impressions have remained, right from the start, that the emphasis has always been on insurance and property protection, and that is fine, but the focus has got to switch, and it's got to switch fast. I believe it has begun to switch, obviously to basic understanding of what goes on in the fire compartment, the understanding of materials, and their interrelationships. Obviously, John Lyons' effort is a key to my survival, but it is also a key to these issues. I think, when we have some better technological answers, we can begin to make the types of tradeoffs that the first speaker addressed. Then we will know how much safety is enough. I think safety is like pornography, just as someone has said, "you know it when you see it." I don't think we even have enough data right now to know what we are looking at. Secondly, one issue which really concerns me is that of how far standards can go. I really don't know how far standards can go or cannot go. If they double John's budget, they could go a lot farther than they would where it stands. Before standards go anywhere, I think there is a world of things that need to be recognized. First of all, I think standards need to focus on end-use and on end-use constraints. That has not occurred in the past, and that is a large part of the problems that I have been concerned about. And I think, secondly, industry has to begin to get itself in a mood (when I say industry I really mean the fire community), which is much broader than private industry, has to begin to put itself in a position to disclose characteristics of products as they begin to understand them. They have to be disclosed to consumers; I mean that in the broadest sense, not to me and my mother. They have to be disclosed to the fire protection people also so that they will know what they are doing when they go into a building. The problem, from my casual observation, is that people fixate very quickly on any given test. I think that a part of the problem with the standards-setting function is that penchant, and it's understandable, but there has to be some greater flexibility because, as I said, the state of the art is evolving. That's what this conference is about. And, as the state of the art evolves, the process has to be flexible enough to evolve with it. It can no longer be fixated, and that goes for standard setters, standard users, and for certifiers as well. Everyone has to become flexible. One other casual observation is that I think building code authorities, never mind the Federal Trade Commission, have to seriously take responsibility and seriously consider themselves as regulators and begin to really flex and to make some hard judgments. I think that there has been a lot of abdication in that area. That should be enough to send three or four people up the wall.

*F. K. Willenbrock*—We are now going to allow equal time for other people who would like to speak on this subject. John has asked to speak.

*J. W. Lyons*—I would like to make a comment about what Eric has said. I was most embarassed to discover, in working with Eric, that I couldn't answer his most basic questions. He wanted to know what reliable, well-grounded scientific test method we had to evaluate the hazard of a product, and, you know, in three years, I haven't answered the question yet. We really don't have such a test. That is embarassing because one could say to the Bureau, you have been doing fire research for 70 years, how come you don't have an answer? It is a difficult problem, but I have appreciated the chance to try to instruct Mr. Rubin and the Federal Trade Commission.

*F. K. Willenbrock*—Are there other panelists who wish to comment? If not, may we have comments or questions from the floor? Would you please identify yourself, confine your remarks to no more than 3 min, and please end with a question?

*J. V. Ryan*[10]—I have a comment parallel to your first question: "How far can fire standards go?" This is not in a fire area but in a product hazard area where we do have some data. One of the first things the Consumer Product Safety Commission addressed was that on safety of bicycles. An analysis made on an in-depth study showed, if I remember correctly, 17 percent of the accidents could be traced to a defect in the bicycle and 83 percent to a defect in the way people, often children, used the bicycle. I choose the bicycle because I don't know enough to give you numbers on the fire area, but this relates to what Joe Clark was saying about nonregulatory approaches and, a little less specifically, to what Armand Burgun was saying about social values. In many cases, a standard can only take us part way because that is all that can be addressed through regulation. My question is, is there anyone up there or out here who disagrees with me seriously?

*F. K. Willenbrock*—Does anyone disagree with Jim Ryan? I see Bob Blake has offered to disagree.

*Robert Blake*—No, I'm going to start off by agreeing with Jim Ryan, and I will relate to you a little scene out in Chicago when the architects and engineers representing the states for the midwest met there at the HEW Regional Engineers Headquarters. The issue was alarm and communication in nursing homes and related regulations. All the architects in the state governments were banging on the Feds (you know that's the way it goes) with regard to the regulation. The air was cleared rather quickly and well, I thought, by the architect from the state of Michigan who sort of swept it aside and said that alarm communication is not necessarily associated with mousetrap devices but that it could rely on some common-sense approaches in training. In a nursing home, this could involve doing some drills, etc., and that would not require reliance on de-

[10]Assistant to the director, Office of Standards Coordination and Appraisal, Consumer Product Safety Commission, Washington, D.C. 20207.

vices which tend to become expensive which they couldn't afford anyway. I think that what he saw was a common-sense approach to a regulation which seemed to require all of those nursing home operators in that area to install bells, blinking lights, etc., to communicate with deaf, blind, etc., people that a fire was going on. This is a long-winded answer to the point you raised, but an important point is that we rely, perhaps too much, on devices. We just love gadgets.

*J. A. Burgun*—May I just respond to that? Perhaps I was too cryptic in the beginning, and maybe I'm pretty much of a skeptic because I've been working with the life safety code since 1963 (not that I think it's a perfect code because it's far, far from it) and yet only a handful of hospitals and nursing homes comply with it. A major reason is that the average age of our hospitals and nursing homes in the United States is increasing. Primarily, we're just adding something more to something that's already there. Even my own buildings don't comply for that very reason. We work at Columbia Presbyterian Medical Center, but we just add a wing here and a wing there, and we don't have the money to make the whole building comply. What has HEW done about it? In 1970, they came along with the Social Security Administration and adopted the Life Safety Code, but they didn't put any money in to correct the deficiencies. All they said was, you either correct them or we will decertify you, and some of the hospitals or nursing homes voluntarily withdrew from the program rather than correct. This is not getting it corrected, and this is exactly the point I was trying to make. You can have an ideal standard, but, until it is incorporated into a building, you haven't accomplished much. And it is that implementation that we must accomplish. Also, HEW has dropped Hill-Burton which has gone out of existence. They don't fund anything anymore.

*F. K. Willenbrock*—Other comments, or next question from the floor?

*H. L. Malhotra*[11]—Can I start by saying that this last week and a half has been extremely useful for me. Both the International Council for Building, Research, Studies, and Documentation meeting and now this symposium have covered a whole range of topics of great interest, ranging from fundamental research to standards and codes. Perhaps because of my involvement with fire tests over the last couple years in connection with the International Standards Organization, I may be permitted to make one or two comments. Particularly, if I may misquote a famous saying, I think the fault lies not in our fire tests, but in us. I think we have misused them a great deal in the past. I think we have expected too much from them, and, when they have not delivered the goods, there has been dissatisfaction and a great urge to throw them away. I don't think this attitude is the right one. We need to undersand the capabilities

---

[11] Fire Research Station, Borehamwood, Herts, United Kingdom.

of fire tests, what information they can provide, and what is the best use we can make of that information. As you wanted me to end with a question, I would like to ask, "How do we arrive at a useful level of protection for our buildings?"

*F. K. Willenbrock*—We have one panelist who agrees with you. Any other comments? John.

*J. W. Lyons*—I think, Bill, I would disagree to some extent with your opening remarks. I don't think all of our fire tests are all that good. Some of them are, and many of our fire tests have a far greater capability than we use. We will make a measurement with almost any test. We will report a number, and we will throw into the wastebasket all sorts of interesting data. For example, consider time dependency. We'll make measurements at 30 s, 1, 2, and 10 min, and then throw the data away and report only the final result. For example, the tunnel test, Test for Surface Burning Characteristics of Building Materials (ASTM E 84-75), has been treated that way in the past. So there's a lot of information there we don't use. On the other hand, some of the tests are an abomination.

*F. K. Willenbrock*—Other comments?

*Wolfram Becker*—May I counter Jim Ryan's statement about the 83 percent drivers who are involved in accidents. The problem is that these drivers are a very bad risk, and they are involved in accidents. If I transfer this to our fire protection problems, are the careless housekeepers and those engaged in setting fires a part of our problem or not? If they are, it seems to me that we will have a real problem. If they are not, then we underestimate the risk, and perhaps this may be a key point, a challenge for the future, to get better figures on fire loss. And if you find out, with your new National Fire Prevention and Control Administration, that this is a key point, then perhaps you can succeed in reducing your fire losses in life and in property by 50 percent in the next generation. If not, what then? Will you turn the screw and ask for more severe fire regulations and so on so that you can't build any building anymore? You can't use any heating device, and so on? What advantages are there in that? This is a big challenge, and we have no answer. Today, what can we do when the people are aware of the risk and are setting fires more or less deliberately? Thank you.

*Larry Seigel*—Karl, along with this matter of getting more information, I'm reminded of a remark that Walter Berl made yesterday along the line that he was suspicious of the 12 000 death figure that has been quoted over and over, and he thought it was something more like 6000. Now, whether he's correct or not, the fact that he suspects this arouses my concern because I certainly have respect for Walter's opinions. If we could prove his 6000 figure, we'd have a 50 percent improvement in fire safety immediately. Being a practical sort, I suggest perhaps we'd better look at those figures.

*F. K. Willenbrock*—Jim Ryan's question reminds me of a story that I heard quite a while ago, before the Consumer Product Safety Commission was in being. The tractor manufacturers were concerned about the safety of farmers using tractors. They decided that one of the problems was that the tractors rolled over too frequently, and the farmers would have greater protection if roll bars were put on. So the manufacturers put roll bars on. But they found that the statistics came out just as they were before, the same number of accidents. It turned out that the farmers were using their tractors to go up higher hills because they had roll bars to protect them. In some ways, you can get beaten at your own game. Are there other questions?

*C. A. Clark*[12]—On the subject of life safety, the discussion just came down to that 12 000 figure again. I think we need to say something about that. I don't know how good that figure is, either, and I also don't know how good the figures were, let us say, in the copy of the *Handbook of Fire Protection* that I have dated 1940. At that particular time and for some years earlier, the fire deaths in the United States were listed at 10 000. From then until now, which is roughly 36 years, we have had, I guess, a doubling of the population in the United States, so, effectively, I think the goals that we talk about and those that the fire community, in the meantime, have been working on have been in the right direction and have shown results. Now, I think that is very good. My question is, will we be able to identify realistic goals as well for the future with the help of the National Fire Incident Reporting System and the Johns Hopkins type of fire death cause studies?

*J. E. Clark*—I would like to make a comment on that. It seems to me that, so often in the Federal government, when the way is cleared for attack on a problem, the priorities, at least in quite a general way, have already been established for you.

Money may become available to work on safe exits from dead-end corridors, when, in fact, that money might be more cost effectively spent on smoke detectors or public education. There is relatively little hard questioning of the basic public policy issue of how we set the priorities for the different attacks that we mount. As resources become available and several years from now we approach that asymptote, as we always do in the funding curve, we will need to worry about which are the top problems. In fact, we have some difficulty in the National Fire Prevention and Control Administration (NFPCA) getting agreement among our management team as to whether or not death should get a higher priority than injury or property loss. It does present a stimulating and important topic for discussion, and it does result in substantially different kinds of programs, depending upon how you establish priorities among those three.

---

[12] Senior development scientist, B. F. Goodrich Chemical Co., Avon Lake, Ohio 44012.

Finally, our inability even to measure those losses accurately and precisely is terrifying.

*J. W. Lyons*—I would like to add to that one point I think is of interest. If you look at the fire statistics for the year 1900, as a member of my staff did, you can estimate that the efforts of the fire community, primarily the efforts of the code officials, the fire services, and really the effect of the fire endurance concept, cut the fire death loss rate, that is deaths per million, by a factor of four from 1900 to this time and the property loss by a factor of two in constant dollars. So the efforts over the last 75 years have been very, very productive indeed, and I would not want anyone to think that we haven't been doing anything. That might have been one of my answers to Eric when he asked me what we have .been doing. We have been doing something but that was generally providing protection from the multiple loss fire in large buildings.

Secondly, if you do look at the retrospective fire fatality statistics, NFPA has details on a very clear category of some 27 percent resulting from smoldering and flaming fires and furnishings at home. I think I indicated yesterday that we estimate some 18 percent of all fire deaths come from the upholstered furniture at the scenario of the smoldering fire. When you have a scenario that is that much different, I think five times as many deaths as in the next nearest scenario, it is clear that that's one place you want to put some priority, and we have. Indeed, we have worked on the upholstered furniture standard at the Bureau for four or five years. And now the policy makers must decide what to do with it. Technical facts are more or less in hand; we can refine them, but they are now in front of the Consumer Product Safety Commission, and they, fortunately, and not I, have the responsibility to make the next decision.

That brings me to my last point, and that is the great danger, I think, of relying too much on fire investigations. We must remember that, when we look at fire statistics, we are always looking backwards. And what has happened in the past may not be what's going to happen in the future. If we change, as we have in recent years, the design of buildings and the kinds of materials we have put in, what earthly good does it do us to look back over the last 50 years. We got into this in a very difficult way in the study in Washington, D.C., of the Metro buses and subways, as some of you are aware. Some people who should know better asserted that the fire statistics will not support an effort on such vehicles. My answer is, of course, that the vehicles in those data bases are mostly very spartan and do not have combustibles. The new vehicles are very plush, with combustible interiors. So don't overrely on data bases. They are always retrospective, and the past is not necessarily prologue.

*F. K. Willenbrock*—Thank you, John. Our time is up. I want to thank our panel members and our alert audience for helping to make this a most successful meeting.

# Summary

The task of summarizing the message presented by a symposium as broadly ranging as this is not an easy one. A technical solution to the fire problem is most elusive and yet we are committed to finding one.

We are, through legislation, in the process of establishing a series of new fire safety requirements for consumer and other commercial products. It seems only reasonable to consider the benefit: cost ratio derived. How can this be achieved if we do not place high priority on some form of loss incidence reporting? However, the discussions in this publication have raised serious questions as to our ability to define the nature and size of our fire casualties even with an accuracy of only about a factor of two. We hear that our insistence on states' rights has resulted in important variations in the content and even the existence of fire loss records. We have difficulty providing adequate incentives to compensate the fire services for their work in complying with a uniform loss reporting system. And yet the need for the resulting information is obviously very great. Thus, the progress reported by Tovey on the establishment of a national fire incident reporting system is welcome news. It seems likely that we will find it necessary to place major emphasis on special studies to yield the information required to help us guide and measure the effects of each particular remedial action we impose as a means for modifying fire loss.

The work reported by Berl and Halpin suggests that the use of alcohol and cigarettes is a major factor leading to many fire fatalities. If further special studies were to confirm this suggestion, would it not be appropriate to question our continued use of such offenders? The question seems to be one of priorities. Perhaps the only reason we have not faced up to it in a serious way is the fact that fire is one of the less frequent and visible fatal consequences of the use of these materials. This, though, is a common obstacle to public appreciation of many fire hazard situations.

The paper presents again the increasing evidence for the importance of the toxic hazard as a basis of fire fatalities. The role of carbon monoxide is stressed, but it is encouraging to see evidence of increased technical competence in detecting the effects of other gases during *post mortem* examinations. This surely is a field in which increased activity deserves support.

The paper by Bercaw et al provides valuable new information on estimating burn severity likely to be associated with fires involving various garment and material types. The manikin model that is being used for research studies is exemplary of a high level of fire system test methods through which the many components and interacting forces of a real fire react to simulate the thermal exposure and thus the burn severity a person might experience. The problem is, of course, to provide a test method, which can be used on a routine basis, which will yield similar data. This paper raises questions concerning adequacy of existing tests used by regulatory agencies and suggests that more attention should be directed toward heat release characteristics. Studies of this problem are currently in progress both in industry and governmental laboratories. It seems evident that, as technical sophistication increases, there is increasing recognition of the problem of providing test methods which cater to the complexities of fire behavior.

Dr. Friedman's paper on the burning behavior of solids is the first of three which provide a review of various aspects of our technical understanding of the fire problem. This is an excellent review of the subject, considering as it does the ignition, smoldering, flame spread, and burning phenonema associated with combustion of solids. In studying this paper, the fire protection engineer may be impatient that wooden and other cellulosic based materials have not been given more attention. The explanation is that, until a competence is developed through understanding the burning of ideal fuels, the research worker becomes frustrated with the more complex problems of cellulosic and other common fuels. The author has surely given evidence of an increasing capability to deal on a technical level with the problems of unwanted fires. The emphasis on the need for further work on smoldering combustion and radiant heat transfer phenomena during fires is well warranted. Clearly we must encourage studies in these areas, as well as support means for applying our present understanding to the fire behavior of the more common materials.

The paper "Some Problem Aspects of Fully Developed Room Fires" by Thomas is a discussion of technical problems faced by a research worker today in his attempt to generalize the theory of fire phenomena. He suggests three problems: (a) can the factors which are decisive in controlling the transition from growing fire to flashover be identified? (b) can theory help us understand ways of controlling and extinguishing fires? and (c) is it likely that fires involving synthetic polymeric materials will result more frequently in explosions than suggested by experience with wood-fueled fires?

The discussion he develops is unconventional and likely to be of greatest value to the advanced fire research worker. He suggests that the approach he has used is oversimplified but indicative of an increased understanding which might develop through further studies. His goals have been outlined but not reached in definitive form.

The review paper by Quintiere, "Growth of Fire in Building Compartments," presents an excellent review of previous work in this field. The data on burning rate of various building furnishing items or components, given in Fig. 2 of his paper, is a measure of the energy release rate of the various fires considered. It presents a most useful summary of presently available information. The main thrust of the paper is, however, a demonstration of achievements which can be made through the use of computer-based fire modeling techniques. By analyzing the various components known to be important in controlling behavior of fires in compartments and assembling these as parts of a computer model, he has exhibited a capability of predicting the general burning characteristics of such fires. This represents a significant advance in fire research. He seems to have properly identified several areas meriting further research work. Important among these is a better understanding of the fluid mechanics of the fire system and of the interaction of flames and smoke on the burning of solid fuels.

Dr. Wakamatsu's paper on prediction of smoke movement in buildings through use of a computer model is illustrative of numerous recent attempts to provide an engineering design system for improved fire safety. The Japanese have been leaders in this activity. It is understood that the techniques described are currently applied in Japan as part of the design and approval activities when large buildings are to be constructed. Following construction of the building, it would be highly desirable to perform air movement tests on the completed building to ensure that both practical performance and building construction features were as proposed when the building design was approved. Such tests should form part of a standard building construction inspection procedure.

The paper by Roux is the first of nine which have been introduced to illustrate the ways in which society has tried to come to grips with the fire problem. It should be considered in relation to the new policy on fire test methods recently adopted by the Board of Directors of the American Society for Testing and Materials (ASTM). This defines a fire hazard standard as one which can be used to measure the fire hazard of a product or system in the form and ambient conditions expected during its normal use. Roux wrestles with this problem in his paper "The Role of Tests in Defining Fire Hazard: A Concept."

A discussion is presented of the importance of recognizing that both a potential for harm (PH) and exposure of people or property (E) terms are involved in an assessment of fire hazard. The resulting search for a rational way of defining these two factors makes clear the problem we all face in providing a technically valid way of evaluating, on an experimental basis, the fire hazard associated with a given product in its expected use environment. Roux proposes the use of a series of separate fire characteristic test results in deriving the PH term. He outlines the currently proposed ASTM requirements for the various possible fire characteristic (FC) tests. But he also refers to the Department of Commerce mattress flammability standard as an FC test. Caution is necessary here. Customary use of fire tests for regulatory purposes has assumed that a single test can be used to define the overall fire behavior of a product. In many cases, this has been accomplished with a fair degree of success. Tests exhibiting this capability should be classified as of the fire system rather than fire characteristic type. Such system tests provide an overall physical and chemical model of some real fire situation. The model may be a good one, but, in many cases, it is very imperfect yet continues to be used as a predictive model. In general, the fire characteristic test usually only measures one of a large number of fire properties of a fire system. The mattress flammability test is a fire system test method.

The two different types of tests must be used in quite different ways, and it seems most important to emphasize this fact. In many instances, the fire property test will yield a physical property of the material or product studied. Such properties can be defined and often measured in several different ways. On the other hand, the fire system test result can only be defined through reference to the test method used. Obviously, in one case, several test methods must be applied to achieve an assessment of hazard, while in the other situation, an overall assessment is achieved from a single test.

The paper, "Building and Fire Codes," emphasizes the importance of better lines of communication between two groups of officials: those charged with regulating building constructional safety and those responsible for safe use of buildings after occupancy. The dual regulatory requirement probably developed as a result of changing occupancy of buildings and the resulting continual need to review its nature. Both types of codes face the common need for ways in which safety can be ensured in the face of continual innovation in type, form, and character of manufacturing processes, products, and materials used for building finish.

The author seems well advised to call for a rational engineering approach to the fire safety problem. It is unlikely that we can afford complete safety or, as suggested by one of the panel members, that we would enjoy it if provided.

The title "A Place for Voluntary Standards" speaks clearly for the theme proposed. It is one of confidence in the voluntary standards development procedures used in this country today. However, it left unanswered several important questions. Is this confidence warranted in the case of fire standards? What incentive exists to ensure the development and maintenance of adequate fire safety standards? Recently, the government, through the Federal Trade Commission and the Consumer Product Safety Commission, has provided a strong stimulus. Does not the very fact that these agencies were formed raise questions concerning the adequacy of the voluntary standardization process? The author concludes with assurance that, if the voluntary standards are adequate and their use satisfactory, there will continue to be a place for such standards. But how do we assure these results; on what basis do we assess the merit and appropriate use of fire standards? Too often in the past fire safety has only been tested at the expense of the public long after initial building occupancy or product purchase. It seems quite clear that the ASTM policy on fire standards, calling as it does for adequate explanatory material on appropriate use and possible misuse of a standard, is a constructive step in forcing a technical evaluation of the merit of such standards.

The Consumer Product Safety Commission is one of the newer federal watch dogs for protection of the public from unreasonable accidents. Thus, the paper "National Mandatory Fire Safety Standards" by Thomas and Willis discusses areas where the government has considered it necessary to establish minimum levels of fire safety. It is expected that the procedures which have been established for processing standards will ensure that technical advice and information available from industry and the public can be considered. While emphasis is placed on unreasonable risks and practical corrective measures, it is not clear that the public is informed on the cost society must pay for increased safety. Further, there is a question as to whether these costs are even known. This is a difficult problem but surely one that must be faced since it is generallly agreed that socio-economic factors may influence fire casualty experience. Too often those most likely to benefit from the remedy seem least able to afford the costs.

The paper "International Standardization" by Witteveen and Twilt stresses the importance of a better degree of uniformity of fire testing as a means for reducing barriers to trade. National borders should not, but often do, influence international trade.

The problem is that fire safety matters are not now subject to exact solution; a comprehensive fire science has not yet appeared. Thus, many fire safety decisions or measurement methods are influenced by subjective and practical decisions. The past history of testing for fire safety, at least in the United States, has resulted in approval of constructions through many costly tests. Naturally, there is reluctance to discard this past experience for the uncertainty resulting from introducing new test methods. This reluctance is emphasized when there is no certainty that the new methods can be shown to yield more technically valid measurements. Thus, change has been slow in revision of fire test methods.

It is not surprising, therefore, that differences exist on an international basis in the requirements expected of building and other elements. It appears that the logical result of this situation is the importance of providing more basic and logical methods of predicting fire safety.

One of the ways society has responded to fire safety problems has been an attempt to apply scientific methods to define fire safety. The paper by Fristrom presenting a consideration of problems associated with the

"Sampling and Analysis of Fire Atmospheres" is evidence of technical concern. Here is an outline of the formidable task of characterizing the gas atmosphere during a fire. Would the measurements resulting from such a complex measurement facility be of value? How would the results be applied?

Perhaps a research study of this type would be useful in characterizing the space, time, and species characteristics of typical fire atmospheres and thus providing a basis for modeling transient conditions and the influence of ambient variables on them. Certainly, a more detailed space and time map of fire gas components is desirable as a basis for encouraging useful reactions which people should exhibit when fire occurs. Perhaps, though, there are other ways to approach this problem through application of thermal-fluid modeling. However, it seems most important that the type of problem faced by this paper be recognized and addressed. Too often we hear a cry of toxic hazard when only one localized gas analysis has been made.

The paper by Petajan et al looks to the toxic hazard of fire gases. This is an area where there has been greatly increased activity recently. It is a field of great technical complexity. Essentially, all unwanted fires result in toxic combustion products, and the response of humans to them is both varied and complex. We can avoid the problem posed by these products only by preventing fires. This is most difficult, especially in a society which will not or cannot afford the luxury of controls on all building furnishing and occupancy items. Thus, this paper suggests the need to recognize that some components of most fire gases present a toxic hazard. It emphasizes the importance of identifying materials which, when burned, release gases which present an unusual or limiting toxic hazard. This seems a most rational position to assume in consideration of this problem. It should permit identification of materials which are likely to present unusual toxic hazards when exposed to fire and thus obviously deserve to be limited in general distribution for public use.

There has, in the past, been a tendency to assume that the identification of the chemical composition of fire gases will provide information useful in estimates of toxic hazard. While, in some cases, this may provide a provisional approach to the problem, it should only be considered as an imperfect stopgap measure. Analytical chemists need to know what they are looking for unless the work is to assume the nature of a research project. This, together with the fact that fire gases may comprise mixtures of many toxic compounds, makes hazard predictions based solely on analytical chemical data very questionable.

Financial support for research on the fire problem has now developed to the stage that we can meet the need for use of animal experiments in assessing information relevant to toxic hazard. This is the approach assumed by Petajan and his co-workers. Here we see the use of a fire system test method in an approach to the measurement of the potential for harm defined by Roux. It has obvious advantages over synthesizing this property from the many fire characteristic measurements which would result from the use of fire characteristic property or analytical measurements.

However, such measurements must be carefully managed if the hazard from fire gases is to be derived directly from the biological measurements. The exposure factor mentioned by Roux, encompassing as it does the modification of fire gas concentration by dilution and the probability of human exposure to the products, must be recognized in derivation of any estimate of fire gas hazard. Obviously, there is a lot of work to be done in this field beyond the many excellent studies now in progress.

The paper by Benjamin, "Prospects for a Compartment Fire Test," exhibits one solution to the problem discussed earlier by Roux, the problem of assessing the overall potential for harm. The procedure proposed is one that has been under development by a task group of ASTM Committee E-39. It differs from the procedure proposed by Roux since it proposes use of a physical model for prediction of potential for harm rather than the integration of individual fire characteristic test results. This involves a fire system test, a model of the real fire in which the various ingredients of the fire situation are incorporated. Emphasis is placed on selection of a relevant scenario, or assumed fire-development chain, together with appropriate selection of criteria of performance.

It seems that this is the type of test which has been called for by the Federal Trade Commission and the Board of ASTM. This is a test in which the various physical and chemical components of a fire are assembled and allowed to interact to yield a result similar to that likely to be experienced when a real fire occurs. It comes close to being a full-scale fire test. There is, however, a danger of oversimplifying the problem. This seems to have been at least partially considered since a range of scenarios are allowed for as models of different types of fires. It seems likely that the fire safety community will be using tests of this type for many years. As these proceedings make clear, there is plenty of work to be done.

*A. F. Robertson*

Senior scientist, Center for Fire Research, National Bureau of Standards, Washington, D.C., editor.

# Index